AT HOME IN THE WORLD

AT HOME IN THE WORLD

California Women and the Postwar Environmental Movement

Kathleen A. Cairns

University of Nebraska Press
LINCOLN

Library of Congress Cataloging-in-Publication Data
Names: Cairns, Kathleen A., 1946– author.
Title: At home in the world: California
women and the postwar environmental
movement / Kathleen A. Cairns.
Description: Lincoln: University of Nebraska Press,
2021. | Includes bibliographical references and index.
Identifiers: LCCN 2020034106
ISBN 9781496207470 (paperback)
ISBN 9781496226211 (epub)
ISBN 9781496226228 (mobi)
ISBN 9781496226235 (pdf)
Subjects: LCSH: Environmentalism—
California—History. | Environmental
justice—California—History. | Women
environmentalists—California—History. |
California—Environmental conditions.
Classification: LCC GE198.C2 C35 2021 |
DDC 363.70092/5209794—dc23
LC record available at https://lccn.loc.gov/2020034106

Set in Sabon Next by Mikala R. Kolander.
Designed by N. Putens.

For Larry

CONTENTS

List of Illustrations   ix

Acknowledgments   xi

Introduction   1

1. "Feminine Warriors": California Women and the Environment   13

2. Saving the San Francisco Bay   45

3. The Dune Lady: Kathleen Goddard Jones   75

4. Saving the Santa Monica Mountains   107

5. Environmental Justice: The Politics of Survival   137

Conclusion   159

Notes   165

Bibliography   185

Index   193

ILLUSTRATIONS

1. Aurelia Harwood   12

2. Sylvia McLaughlin, Catherine "Kay" Kerr, and Esther Gulick   44

3. Kathleen Goddard Jones   74

4. Jill Swift, Susan Nelson, and Margot Feuer   106

5. Marie Harrison, San Francisco activist and
   member of Greenaction   136

ACKNOWLEDGMENTS

Writing about the place my family has called home since the 1880s has led me to reflect on the many people who have accompanied me on this extraordinary journey through life in the Golden State. Like the women in this book, at the beginning I never could have imagined where the path would lead. I am profoundly grateful to my fellow travelers. In the smoggy San Gabriel Valley, where I spent my youth, I thank Ann Banks Dodson, Mary Ann Prosser Zukowski, and Karen Ballew Theetge, along with many other fellow Glendora High School grads. In Long Beach, where I attended college and spent twelve years as a newspaper reporter—the most fun job I ever had—Susan Pack, Rich Archbold, Mark Gladstone, Dennis McDougal, Jim Nolan, Mike Schwartz, Leo Hetzel, John Futch, and Joyce Christensen. I met Dorothy Korber when we were both young reporters, and our friendship has only grown deeper over the decades, with long conversations about the world, our families, books, journalism, and many other topics. We've hoisted numerous glasses (bottles) of wine and eaten amazing food during visits to her cozy Craftsman home in the mountains. Also, thanks to the late Linda

Kaplan, my partner in "crime," who could always make me laugh, even in the darkest times.

In Davis and Sacramento, where I finished graduate school and got my first real teaching job, I thank fellow members of the Sisterhood: Vicki Ruiz, Olivia Martinez, Alicia Rodriguez, Annette Reed, Margaret Jacobs, and Yolanda Calderon Wallace. Also, Kathy Olmsted, Bill Ainsworth, Janet Freitag, Matt and Nancy Kuzins, Kathy Noonan, Sue Pearson, and George Craft. On the beautiful Central Coast, where I live now, I thank Karen McLaurin and Mike Cussen; Anne and Rod Bell; my fellow bookies, Kathy Myers, Janis Bailey, Derinda Gates, Nancy Peck, Paula Adrian, Valerie Wormuth, Pam Pesenti, Janyce Smith, Pam Young, Fran Long, Carla Crane, Susan Farley, Valorie Marshall, Marilyn Tackitt, Kathy Howard, Ethel St. John, Pat Lapp, Cindy Peyton, JoAnn Peterson; my walking partner, Vicky Hoffman—plus her stellar pup Lucy; the peerless Patti Kane; Ruth Rusch and Jim Pope. Also, my generous colleagues at Cal Poly–San Luis Obispo.

Donna Schuele has been a great friend, close reader of manuscripts, generous sharer of ideas, and conference-going partner. We've traveled to Denver, Minneapolis, San Francisco, and other locations but, sadly, not to Paris! For nearly a decade, Bridget Barry and her phenomenal colleagues at the University of Nebraska Press have guided my manuscripts from inspiration to publication. Bridget is brilliant, unfailingly patient, and kind; no one could ask for a better editor.

Finally, there is my family, without whom I could not have taken this journey: Allison Hungerford, Alexandra LaChapelle, Dave Hungerford, Tom LaChapelle; Katherine Lynch Swain, Anna Anderson Cook, and Al Anderson; the Fab Five—Bailey, Andrew, and Finley Hungerford and Sadie and Josie LaChapelle—Ann and Doug Cairns; Dennis and Carol Lynch. Plus, my fabulous Stella, a rescue mutt who shares my life and has my heart.

During the writing of this book, Lawrence Lynch, my husband, life partner, and best friend, died after several years of serious health problems. I first met him in 1978, when a PSA jetliner crashed in San Diego. He was city editor of my newspaper. The star reporters were sent to the scene of the crash to do the big stories or assigned to do color. The rest of us were handed slips of paper and assigned to get photos of the victims. Mine had lived on Palos Verdes

Peninsula. It was dark and I got lost. Plus, I didn't really want to knock on the door of a crash victim. I found a pay phone outside of a liquor store and called the city desk. Larry answered and listened in silence as I explained my predicament. "Get back in the car and get the photo!" were the first words he ever spoke to me. It was a preview of coming attractions. For nearly forty years he pushed me far beyond what I had long thought were my limits. He believed in me when I didn't believe in myself. He was fearless, patient to a fault, empathetic, and a world-class listener, and he took everything in stride, even illness and the prospect of death. His motto was "It is what it is." This book is dedicated to him. For the record, I got the photo.

AT HOME IN THE WORLD

INTRODUCTION

In January 1961 Esther Gulick, Catherine "Kay" Kerr, and Sylvia McLaughlin decided that something needed to be done about the San Francisco Bay. In the century since California's statehood, the Bay had become the receptacle for mountains of garbage from businesses, developers, towns, and cities. The women met at Gulick's home in Berkeley and created the Save the San Francisco Bay Association. Virtually everyone at the time viewed the three women as naïve, idealistic, and "starry-eyed bay savers." Few could have imagined that individuals so sheltered and so unfamiliar with how politics worked could pull off such a massive endeavor. Yet a decade later Save the Bay had thousands of members, had largely stopped the destruction, and had convinced lawmakers to create a state agency to monitor the nearly five-hundred-square-mile bay, to restore wetlands, and to carve out miles of parks, bike trails, and hiking paths.

More than two hundred miles to the south, Kathleen Goddard Jones mounted a campaign against Pacific Gas and Electric Company, which sought to build a nuclear power plant on the pristine, fragile, and ecologically important Nipomo Dunes in southern San Luis Obispo County. Proponents wanted the tax revenue and jobs such a facility would bring. Goddard Jones plowed ahead, despite the opposition. By 1966 she had convinced the investor-owned utility to abandon the site and build its facility twenty-five miles to the north, in Diablo Canyon. The Diablo campaign precipitated a civil war among members of the Sierra Club, the nation's preeminent conservation organization. But in the end Goddard Jones saved the dunes.

Meanwhile, in sprawling Los Angeles, women became involved in several environmental campaigns. One group, tired of official inaction regarding smog, in 1959 created Stamp Out Smog (SOS), an organization devoted to lobbying state and local lawmakers for tougher regulations on air pollution. They attended public meetings wearing gas masks and presented officials with a smog birthday cake covered in black frosting. Women from another group, Save the Santa Monica Mountains, Parks, and Seashore, fought a nearly two-decade battle to stop corporations and developers from carving up the hills to build apartments, condominiums, subdivisions, and freeways. By the late 1970s they had succeeded in setting aside more than 150,000 acres for the Santa Monica Mountains National Recreational Area.

Activism on behalf of the environment was not a new phenomenon in the years following World War II in California, though it might have seemed that way to many people. In fact, the state had stood at the center of conservation efforts since the naturalist and writer John Muir first stepped foot in Yosemite Valley in the 1860s and vowed to save the glorious landscape from loggers, ranchers, and miners. In 1892 Muir and nearly two hundred friends and acquaintances, nearly all men, created the Sierra Club, headquartered in San Francisco. Initially an organization for hiking and pack-trip aficionados, by the 1960s it was the nation's premier conservation organization, leading campaigns to stop construction of dams in national parks and ski resorts in pristine mountain areas.

Men had long headed the Sierra Club and other organizations. Women had participated in conservation efforts from early days as well, though they received scant attention. They were writers, scientists, mountain climbers, and women's club members. Mary Austin wrote of the state's deserts and mountain regions and fought Los Angeles's theft of water from Owens Valley. Alice Eastwood worked as curator of the San Francisco Academy of Science and traveled the length and breadth of California, looking for specimens to add to the academy's collection. Marjory Bridge Farquhar was one of California's, and the nation's, premier mountain climbers; being in nature, she said, helped foster a dedication to preserving natural resources. Laura Lyon White and the California Federation of Women's Clubs worked to save the stately coastal redwoods from loggers.

By the 1950s, however, new issues and problems had emerged, most attributable to the massive and seemingly uncontrollable growth industry that exploded in the years after World War II. By 1960 California was closing in on New York as America's most populous state. Seemingly overnight, suburban neighborhoods, shopping centers, and schools emerged from thousands of acres of plowed-under groves of trees and hundreds of thousands of acres of agricultural land. Freeway builders brought more problems. They dug up streets, razed neighborhoods, and fueled an increased reliance on cars. While freeways proliferated in Southern California, residents in the San Francisco Bay area mounted a ferocious resistance. But developers there built entire towns on landfill that resulted from years of dumping waste products. And utility companies promoted nuclear power plants up and down the state. They would provide cheap and clean energy, or so the argument went.

New problems required new strategies and voices. Mired in the past, many male leaders seemed paralyzed, uncertain how to confront the myriad challenges. Women stepped into the breach, in the process moving beyond established organizations to forge a new kind of grassroots community activism. It is not too far-fetched to view them as midwives of modern environmentalism, in some cases working in opposition to or with limited assistance—at least initially—from powerful men.

Investigating how a long-standing focus on preserving natural resources in wilderness areas morphed into a much broader environmental movement aimed at communities where people actually lived is itself an important endeavor. But women, not men, fueled this transition. Initially derided as "housewives," and with virtually no access to political institutions or the male power elite, they slowly emerged into public view. They attended hearings, where they buttonholed politicians and demanded to be put on the agenda. They published newspaper articles and made themselves available to reporters. They mounted public relations campaigns targeting recalcitrant officials. Their successes earned them first grudging then outright respect from politicians and from their male counterparts. Most important, they got results, which brought additional recruits to environmentalism and ultimately turned California into a leader in the burgeoning environmental movement.

Interestingly, many of the female activists lived very long lives for the time, suggesting that activism was good for the spirit and the soul. Two of the three principals of Save the San Francisco Bay—Kerr and McLaughlin—each lived to the age of ninety-nine. Goddard Jones lived to the age of ninety-four; she was still hiking the Nipomo Dunes in her early nineties. In Alaska, Margaret Murie, an activist on behalf of the Arctic National Wildlife Refuge, lived to the age of 102; and in Florida, Marjory Stoneman Douglas, who worked to save the Everglades, lived to be 108.

Rachel Carson is often cited as the woman most responsible for facilitating the modern environmental movement. A scientist and writer, her 1962 book, *Silent Spring*, represented a clarion call against dangerous and often lethal pesticides. Carson's work was extraordinarily significant. It encouraged public skepticism about the relationship among science, business, and government, one based on profit rather than objective research. But some of the women who worked on environmental issues, in California and elsewhere, became activists prior to the publication of *Silent Spring*. And they created community organizations dedicated to effecting change.

Before they became politically engaged, their lives seemed to reflect the prevailing post–World War II ideology that sold white, middle-class domesticity via books such as Dr. Benjamin Spock's *Baby and Child Care*, in family-centric television programs such as *Leave It to Beaver* and *Father Knows Best*, and in the seemingly endless rounds of carpools, Little League games, dance and piano recitals. High school girls during this era were pushed toward typing and shorthand classes rather than math or science, and, if they attended college, toward home economics as perhaps the single most useful major for female students. Elementary school teaching topped the charts as the career of choice for many female college graduates, who, after a few years, were supposed to quietly return home to marry and await the births of their children.

Beyond surface appearances, however, the women's lives were much more complex than those depicted in popular culture or in Betty Friedan's seminal 1963 book, *The Feminine Mystique*, which emphasized homemakers' alienation, anxiety, and isolation. They graduated from college in numbers far higher than the norm for the time, many from prestigious universities. They did not major in home economics. Several majored in journalism. They included Harriett Weaver of the Federation of Hillside and Canyon Associations in Los Angeles, at the University of Kansas; and Dorothy Erskine, a founder of People for Open Space in San Francisco, at the University of California, Berkeley. Kay Kerr of Save the Bay majored in political science at Stanford. Her colleague Sylvia McLaughlin majored in French at Vassar. Margot Feuer, cofounder of SOS and active in Save the Santa Monica Mountains and Friends of the Santa Monica Mountains, Parks and Seashore, majored in music at Wellesley. Kathleen Goddard Jones attended Mills College in Oakland; she planned to be a teacher but dropped out to marry an Iranian pilot, whom she later divorced. A few women earned advanced degrees. Sue Nelson, a leader in the Santa Monica Mountains campaign, held a master's degree from UCLA in political science. Claire Dedrick, active in a variety of Bay Area environmental organizations, held a PhD in microbiology from Stanford.

Many of the women also traveled widely, both before and during marriage, so they knew something about the world beyond their neighborhoods and communities. McLaughlin traveled through Europe with her family as a child and spent time in France as a college student. At the age of nineteen, Goddard Jones spent three months in Europe, mostly on her own; in her twenties she lived in Iran, India, Burma, and New York. Erskine became interested in communism, so she traveled twice to the Soviet Union in her twenties to see the system operating firsthand. Gulick traveled to Europe and Asia frequently throughout her life.

They belonged to a wide array of community organizations, so they knew something about how to work with other people on various causes and campaigns. Some of the groups focused on conservation. Goddard Jones and McLaughlin belonged to the Sierra Club, as did Feuer. Goddard Jones also belonged to the Audubon Club, and she started a local Native Plant Society chapter.

Many belonged to the League of Women Voters, an organization created at the end of the women's suffrage movement and designed to teach women how, as equal citizens, to navigate and utilize the political system. Though larger social changes did not come until later, they were the kind of women who might have paid attention to the Commission on the Status of Women, created in December 1961 by President John F. Kennedy at the behest of Eleanor Roosevelt to investigate and address discriminatory laws and practices relating to women in government and in employment.

Virtually all of the white female California environmental activists in the 1960s were married to prominent and successful men, which undeniably helped them gain access to members of the political establishment. And their spouses' success meant they had money, household help, and time to pursue causes about which they were passionate. Caroline Livermore was president of the Marin Conservation League; her husband, Norman, came from a wealthy family and was a founding board member of Pacific Gas & Electric (PG&E). Kerr's husband, Clark, was president of the University of California system. McLaughlin's husband, Don, owned a mining company, and Gulick's husband, Charles, taught economics at UC Berkeley. Erskine's husband, Morse, was a prominent San Francisco attorney. Goddard Jones's second husband, Duncan Jackson, owned olive and almond groves on California's Central Coast and sold the almonds to the Hershey Company for its Almond Joy chocolate bars. Weaver's husband, John, was a successful writer and screenwriter. Some women came from prominent political families themselves. Betty Brown Dearing of the Federation of Hillside and Canyon Associations was the daughter of one Ohio congressman, Clarence Brown, and the sister of another, Clarence Brown Jr.

Most of their spouses wholeheartedly supported their activism, and some offered professional assistance. For example, Morse Erskine provided free legal advice and services to a variety of organizations in which his wife was involved. This support facilitated the women's work because it helped to defuse potential criticism that they were abandoning their domestic roles. John Weaver was possibly the most supportive husband of the lot; he wrote articles and even books extolling Harriett as an extraordinary person virtually worthy of sainthood. After one successful campaign to stop bulldozers from digging up the hillside across from their Los Angeles home, John exulted, "Harriett refused to accept the cynical notion that nothing could be done.... She made her last stand before the [Los Angeles] City Council, where the enemy was done in by a unanimous vote. Next day she watched the bulldozers slink down the hill in a whimper of steel."[1]

Goddard Jones's husband was possibly the least enthusiastic, if only because the couple lived in a small town and he feared her activism would hurt his business. Yet even he offered some support, helping his wife organize Sierra Club gatherings and traveling with her to examine the prototype of a nuclear power plant. He drew the line, however, at midnight phone calls from antagonists, some of which were threatening in nature. "Is it smart to go against everyone?" he once asked. Since she cared little about criticism, and he cared a lot, Goddard Jones became one of the few women activists to be divorced—in her case, twice.[2]

Because they were engaged in somewhat untraditional enterprises for their time, the women often framed their work as domestic in nature; after all, it dealt with issues that affected communities, families, and children. Most also understood that they needed to look traditional during appearances before official bodies, when they gave speeches, and when they participated in public gatherings. In oral histories, books, and magazine articles, they described dressing with great care. Goddard Jones was always much more comfortable in hiking clothes and boots, with a tin water cup hanging from the waist of her pants. Yet she wore blouses, skirts, high-heeled shoes, gloves, and hats to meetings. McLaughlin always wore jackets, skirts, and medium-heel pumps. High heels were too uncomfortable; flat shoes might remind her audience of "little old ladies in tennis shoes," she feared.

And they labored to appear in control at all times. More than one described hands trembling as they spoke to gatherings or of jamming their hands into their pockets so observers wouldn't see them shake. Clearly they were being judged. After one particularly tense battle over hillside development, a Los Angeles City Council member congratulated Federation of Hillside and Canyons representative Betty Dearing for her "considerate and ladylike behavior," adding, "Certainly, your logical and forthright approach had much to do with the passage of my resolution."[3]

It is impossible to discuss the activist women without noting that their lives and campaigns appeared elitist to many observers. And understandably so. They were economically privileged, socialized with few individuals outside of their own class, and rarely had occasion to interact at all with those less fortunate unless they were employees—their own and other people's. As a result, they could be tone deaf or oblivious. For example, the Save the San Francisco Bay Association grew so rapidly in its first few years that its leaders needed to hire part-time staff. They hired Janice Kittredge, who worked at UC Berkeley. As Kittredge recalled later, neither Gulick nor Kerr seemed aware that she needed a steady paycheck.

If confronted with allegations of elitism, the women expressed dismay; they were working to make the environment better for everyone, they insisted. It was difficult to argue that hiking trails and public parks around San Francisco Bay or clearer skies in Los Angeles were elitist endeavors. But laboring to keep multifamily dwellings out of the Santa Monica Mountains could be deemed self-interested and exclusionary; the women were worried about their property values, critics declared.

In fact, leaders of the burgeoning movement generally evinced little interest in serious problems in poor or minority neighborhoods. For example, while Rachel Carson led the charge against pesticides, similar efforts by labor leaders, including Dolores Huerta of the United Farm Workers, mostly fell on deaf hears. Huerta and others argued that pesticides caused serious health issues for people who worked in fields saturated with the dangerous chemicals. When many pesticides eventually were banned, the focus was largely on consumers who ate infected food, or even on birds and animals, rather than on farm-workers. And while wealthier Los Angeles residents were busy joining the Federation of Hillside and Canyon Associations to protect open spaces, a few miles away bulldozers razed homes in the mostly Mexican American neighborhood of Chavez Ravine to make way for a stadium for the newly arrived Los Angeles Dodgers. No federation member sought to stop the bulldozing.

Those evicted faced limited choices when it came to finding replacement housing, since racial segregation was still the norm in many, if not most neighborhoods in California and elsewhere. There were restrictive covenants even after the U.S. Supreme Court outlawed them in 1948. And proliferating homeowners' associations made private property rights paramount. Most white, middle-class Californians were not prepared to change that reality.

In fact, "elitism" had been a label long applied to conservationists. In the nineteenth and early twentieth centuries the insistence of the Sierra Club and other organizations on keeping wilderness pristine angered those who lived and worked there. And the movement had long skewed white and middle to upper class. From its early days the Sierra Club required prospective members to obtain sponsorships from members in order to join the organization. But even having sponsors did not always ensure acceptance. An African American woman was denied membership In Los Angeles in the 1950s despite having the required sponsors and the fact that she was an "ardent and experienced hiker and car camper." The Club began to abandon the sponsorship requirement in the 1960s, however, as leaders embarked on extensive recruiting efforts.[4]

When the post–World War II housewives became activists, this societal shift had not yet occurred. Virtually all middle-aged, they came to adulthood at a time when few people, particularly in their circles, questioned the status quo with regard to class, gender, and ethnicity. Thus their inability to reach outside their own life experiences in this regard is not particularly surprising. It would be left to future generations comprised of many different individuals and groups to create a broader and more inclusive movement that came to be called "environmental justice."

Like its white, middle-class counterpart, the environmental justice movement has largely been led by women, most of whom had no prior experience as organizers. Groups have included Mothers of East L.A., who fought against a proposed prison and incinerator in their neighborhood; Greenaction in the San Francisco Bay Area, which pushed for neighborhood toxic cleanup; and Citizens for a Better Environment in Los Angeles, which lobbied for, among other things, the removal of a five-story pile of concrete rubble left behind from a 1994 earthquake. In the small, working-class Riverside County town of Glen Avon, activists fought to compel corporations and government agencies to clean up millions of gallons of toxic chemicals from the Stringfellow Acid Pit after a series of rainstorms caused the pit to overflow, spewing toxic waste into schoolyards, gardens, and wells.

Looking back from a distance of more than sixty years, it seems clear that the privileged women who shaped a new focus on the environment had a relatively limited agenda, but they set the movement on its course, and for that they deserve much credit. At the beginning they were political neophytes. Initially focused on ecological threats to their own communities, their work helped to lay the foundation for an ever-expanding agenda that has come to include social justice, a crusade led by individuals and groups with virtually no access to power but an intense dedication to ensuring that their own families and communities have the same access to safe, clean, and toxic-free living spaces as the residents in more affluent areas all those decades ago.

This book has five chapters. The first traces the trajectory of women as conservationists and environmentalists throughout much of California's history. The second traces the campaign to save the San Francisco Bay through the experience of Sylvia McLaughlin. The third focuses on Kathleen Goddard Jones and her crusade to save the Nipomo Dunes. The fourth features a variety of women who labored to limit development in the Santa Monica Mountains. The fifth examines women in the environmental justice movement who significantly broadened the scope of activism to include communities of color and working-class communities.

FIG. 1. Aurelia Harwood was the Sierra Club's first female president, elected in 1927. Following her death in 1928, the Club's Los Angeles chapter named a Southern California mountain after her. Photo courtesy of the Sierra Club.

"FEMININE WARRIORS"
California Women and the Environment

In January 1959 nine women visited the office of Los Angeles County Supervisor Warren Dorn to talk about smog. A *Los Angeles Times* reporter on hand for the meeting dubbed the participants "self-described 'amateurs' in politics." Though neophytes, they were "willing to learn," he acknowledged. All of the women were members of a newly formed group called SOS, for Stamp Out Smog, and they were visiting Dorn "to ally themselves with the county in its air pollution war."[1] A modest and tastefully attired group, all wore dark dresses and high heels; some wore hats or pearl necklaces and clutched small handbags. They posed for the camera with legs crossed demurely at the ankles. The cutline accompanying the photo listed all of the women by the names of their husbands, many of whom were well known locally, if not nationally.

There was Mrs. Robert Cummings and Mrs. Art Linkletter, both wives of noted entertainers. There was Mrs. Jerome Zeitman, whose husband was vice president of Music Corporation of America; Mrs. William Orr, married to a vice president of Warner Brothers studio; and Mrs. Clarence Shoop, married to a military general who worked at Hughes Aircraft Company. The reporter described the assembled women as "feminine warriors" for a good cause: cleaning up the choking, filthy air that blanketed the Los Angeles basin for up to six months of every year.

The campaign against smog was not new in Southern California. For more than a decade, Angelenos had confronted the problem that first emerged during the region's industrial buildup during World War II. Since Los Angeles County sits in a basin surrounded on three sides by mountains, ocean-fed breezes flowing from the south and west pushed the fetid air toward the interior hills, where it remained trapped until fall and winter rains temporarily cleared the skies. Over the years, officials created an Air Pollution Control Board, which issued smog alerts. They commissioned studies to unearth the causes of smog. Private citizens led letter-writing campaigns and held dozens of community meetings. Who was to blame? Was it automakers? Industry? Backyard incinerators? Men had led the policy discussions and meetings, with few tangible accomplishments, except for the banning of incinerators, used mostly in poorer neighborhoods.

Enter the women of SOS. They had grown weary of debates, discussions, and arguments. Meanwhile their children attended schools that canceled sporting events and sent them home when the choking air made it too hard to breathe. Parents were warned to keep their kids inside during the worst days of summer and early fall. Gathering in the living rooms of their comfortable homes, the women plotted strategy that leaned heavily on garnering media attention for an all-out campaign to force elected officials into action. Their meeting with Supervisor Dorn represented a first step. Generally considered conservative and pro-development, Dorn had made the fight against air pollution his personal crusade.[2]

Implicit in SOS's strategy was the women's understanding that they needed the tutelage and support of powerful men to successfully launch their political journey. Their body language and style of dress reflected this realization; they clearly understood the rules. Men made the decisions; women were supplicants in the larger world outside of their homes. They were white, upper-middle-class, even wealthy housewives who possessed social and some economic capital. But they had no real power apart from that conferred by their marriages to powerful men.

They did understand something of how politics worked, however. Many if not most SOS members belonged to other women's groups—garden clubs, the American Association of University Women, the PTA, and the League of Women Voters—so they knew how to organize. The League was particularly significant. From its inception in the years following women's suffrage, its goal had been to help women carry out the responsibilities of citizenship by learning how public policies were crafted and implemented. Their experiences had taught them how to craft a narrative to enhance the impact: smog impacted families and children; it made people sick, and it might even kill them. Thus they could present themselves as striving to help larger society and as engaged in important, even crucial work on society's behalf.[3]

Once they plunged in, SOS members proved to be fast learners. In May 1959 they announced a petition drive to "bombard" Los Angeles County supervisors with letters urging a requirement that "gasoline companies keep smog-producing elements in all gas sold at 12.5 percent or less." By that October they had "launched a statewide offensive . . . aimed at persuading Gov. [Edmund G.] Brown to call a special legislative session on smog control." Toward that end, they contacted eleven mayors in the greater Los Angeles area, seeking the formation of regional air pollution control districts, and they created an alliance with nearly five hundred other women throughout the state.[4]

By 1964 they had achieved some successes. The state Motor Vehicle Pollution Control Board had mandated smog-control devices for cars, trucks, and buses. Only one device had been authorized, however; another still awaited approval before the devices could be installed. Afton Slade, president of SOS and a Beverly Hills resident, announced a gathering to "commemorate" the twenty-first "birthday" of the first smog alert. Before a roomful of reporters and photographers, she unveiled a black birthday cake, decorated with a skull and crossbones. "Smog may win and we may lose the battle for clean air unless action is taken immediately," Slade declared. She urged the Pollution Control Board to approve the second device, and quickly. "What if Dr. Salk had to wait for two [polio] vaccines to be approved?" she asked. "How many lives would have been lost?"[5]

Divorced and the mother of four children when she joined the movement, in 1965 Slade voiced frustration with "the sexist and patronizing" way some members of the media and the public depicted members of SOS. "The picture of the little old lady draped in gray veils, whisking in and out of the hearing room is misleading," she wrote. Her outspokenness did not harm her reputation, or her cause, however. When President Lyndon Johnson signed the Air Quality Act in 1967, he invited Slade to the White House. And Governor Ronald Reagan of California appointed Slade to an air pollution panel in the late 1960s.[6]

SOS was one of a number of environmental campaigns throughout California led by women beginning in the late 1950s. Like the members of SOS, most resembled women featured in magazine articles and in books such as Betty Friedan's iconic *Feminine Mystique*, which detailed the ennui and sense of purposelessness experienced by many white suburban housewives at the time. Most were married and at least middle class, and few worked outside the home in paying jobs. If they suffered from Friedan's "problem with no name," however, they addressed it by looking outward rather than inward. "For a significant minority of middle class women in the United States . . . the pull of domesticity paled in the face of the urgency they felt to address the country's social and political problems," writes Susan Lynn. "Furthermore, domestic values were often employed to support women's public efforts."[7]

In fact the same ideology that limited white middle-class women propelled them toward activism in post–World War II California. Encouraged to stay home and rear children, they took their prodigious energy and intellect into the larger community, recognizing that emerging problems created challenges far beyond the capabilities and skill set of public officials and established organizations. Californians had long grown accustomed to explosive growth, but from the mid-1940s through the 1960s the state experienced an unprecedented level of demand for homes, cars, and consumer products.

During the war thousands of military personnel and defense workers descended on the state that received $35 billion (one-tenth) of the national defense budget. Many war workers remained in the state after the conflict ended; hundreds of thousands of others came to work in the burgeoning aerospace and subsidiary industries in the early Cold War period. The GI Bill drew veterans to California for a virtually free education at world-class institutions of higher learning, including UCLA, Stanford University, and the University of California, Berkeley. Young people whose lives had been put on hold by the Depression and the war married and began having babies—the vanguard of what would come to be called "the baby boom." Between 1950 and 1970 California's population virtually doubled, from 10.5 million to 20 million.[8]

California had experienced a severe housing shortage during the war. Afterward pent-up demand fueled a frenzy of homebuilding, aided by millions of dollars from the federal government, which provided low-interest loans and no down payments for veterans. Developers took advantage of government largesse, buying up millions of acres in both Southern California and the San Francisco Bay Area. They plowed under fields of beans, strawberries, broccoli, lettuce, and wheat and groves of avocado, almond, citrus, walnut, apricot, plum, and cherry trees. By the mid-1950s the former fields held sprawling subdivisions with streets laid out mostly in grid patterns. The houses, produced in assembly-line fashion, seemed largely interchangeable: single-story, stucco, with two or three bedrooms and usually one bathroom, selling for $8,000 to $10,000. "More than anyplace else, California became the symbol of the postwar suburban culture," writes Kenneth T. Jackson.[9]

The San Gabriel Valley, east of Los Angeles, held the bedroom communities of West Covina, Arcadia, and Glendora. Farther south Orange County had Westminster, Huntington Beach, and Anaheim. The San Fernando Valley, north of Los Angeles, had long been ranch land, but by the 1960s "most of the Valley had been transformed into a mosaic of residential communities."[10] The peninsula running south from San Francisco through Santa Clara County once supplied "between one-third and one-half of the world's prune supply."[11] In the postwar period prunes were replaced by houses and businesses. San Jose held a population of less than 70,000 in 1940; by the 1970s, it had become the nation's largest suburb, with more than 600,000 residents. In the East Bay, across the Bay Bridge from San Francisco, burgeoning suburbs included Concord, Danville, and Walnut Creek.[12]

Southern California also had Lakewood. Just north of the port city of Long Beach, it was California's first planned city, built on the model of Levittown on New York's Long Island. Developers laid out streets and began building in 1949: eight houses per acre, seven models of two- and three-bedroom, one-bathroom homes. "One man with a pneumatic hammer nailed subfloors on five houses a day," writes D. J. Waldie, who grew up in Lakewood. "Rough plaster laid by one crew was smoothed a few minutes later by another."[13] When the sales office opened in April 1950, twenty-five thousand people waited in line, ready to pick out the desired model, complete with modern appliances and even a tree for the front yard. They were buying not just a home but what they hoped would be an upwardly mobile, middle-class lifestyle, complete with schools, parks, baseball diamonds, and readily available shopping at the Lakewood Mall. Anchored by the department store May Company, the mall had parking spaces for ten thousand cars. By 1954 Lakewood had a population of fifty-seven thousand—mostly young families. Virtually all the residents were white; segregation, both formal and informal, kept nonwhites out of suburban areas.

Suburbs increased the demand for cars, since they generally were located miles from urban centers were many people worked. Suburbia, as Jackson notes, "is a creature of the automobile and could not exist without it." More cars meant a critical need for roads. The 1956 Interstate Highway Act provided billions to underwrite construction of more than forty thousand miles of roadways—primarily freeways—throughout the country. Devoid of traffic signals and with off-ramps often a mile or more apart, drivers—theoretically, at least—could get to their destinations faster. The 1950s and 1960s saw virtually unceasing freeway construction. Southern California had the Santa Ana, San Diego, Santa Monica, San Bernardino, Hollywood, Long Beach, San Gabriel freeways. Northern California mostly utilized numbers: the I-80, the 280, the 680, the 580.

By the 1970s freeways had become ubiquitous in Southern California. In her novel *Play It as It Lays*, Joan Didion features a lonely, jaded woman named Maria who assuages her depression by driving: "Once she was on the freeway and had maneuvered her way to the fast lane, she turned on the radio at high volume and she drove. She drove the San Diego to the Harbor, the Harbor up to the Hollywood to the Golden State, the Santa Monica, the Santa Ana, the Pasadena, the Ventura. She drove it as a river man runs a river."[14]

But freeways proved to be no panacea; in fact they contributed mightily to urban sprawl and pollution. Families now needed two cars: one for the breadwinner, usually male, who commuted to work, and one for the stay-at-home partner who needed an automobile for ferrying kids and doing errands. Entire industries catered to the car culture, including fast food. Richard and Maurice "Mac" McDonald inaugurated their "speedy service" in a single San Bernardino restaurant in the late 1940s. By the mid-1950s McDonald's was a thriving hamburger chain, with locations throughout California. A decade later it had expanded across the country, on its way to becoming a worldwide phenomenon.[15] Other fast-food chains soon followed, including In-and-Out, also headquartered in California. In July 1955 Walt Disney opened Disneyland, built on the ruins of bean fields and orange groves in Anaheim. On opening day, freeway traffic backed up seven miles on the Harbor Boulevard off-ramp of the Santa Ana Freeway. Five months later one million people had visited the park. Major League Baseball also came to California in the 1950s, with the arrival of the Los Angeles Dodgers and the San Francisco Giants. By the end of the 1960s the state could boast five teams. With limited public transportation, people had to drive to eat fast food, spend the day at Disneyland, or attend ballgames.[16]

To many if not most members of the white middle class in California, the benefits of suburbs, freeways, and other accoutrements of postwar living outweighed the disadvantages. But the cost of modernity was becoming apparent as well; smog and traffic and the loss of open space. Overbuilding on hillsides created mudslides; for instance, in January 1952 a drenching rainstorm sent a "half-million cubic yards of mud and debris" onto Los Angeles streets; "almost two hundred tons of it slumped down the steep side of Beverly Glen Gulch" in the Santa Monica Mountains. While digging out, residents noticed that the worst damage occurred to homes sitting beneath new construction.[17]

It turned out that modern life required new places to dispose of mountains of waste from construction, manufacturing, chemicals, and consumer products, including millions of discarded toys from children of the massive baby boom generation. San Francisco Bay was a depository for many businesses and cities. There were existential issues as well: the threat of nuclear annihilation and fallout from nuclear testing, which engendered a pervasive sense of insecurity and dread. And federal and state officials were selling new ways to harness and utilize nuclear energy: power plants designed to provide electricity to homes and businesses. California, with its 1,100-mile coastline, was ground zero for this effort since power plants needed water for cooling purposes. Was nuclear power safe? What about radioactive waste? Where would it go? No one seemed to have any answers, at least ones they were willing to share.

Public officials were not the only ones seemingly out of their depth when it came to confronting issues and problems associated with postwar developments. Long-standing conservation organizations such as the Sierra Club and Save the Redwoods League, both based in California, had lobbied on behalf of wilderness preservation. In fact the naturalist, writer, and Sierra Club founder John Muir led the Club's first political campaign during an earlier growth spurt. In the early twentieth century it tried unsuccessfully to stop the destruction of Yosemite's Hetch Hetchy Valley to provide water to San Francisco. California remained the Club's focus until the 1950s, when it mounted a successful effort to stop the building of a dam in Utah's Dinosaur National Park. But Club leadership proved reluctant or unwilling to tackle concerns outside of preserving wilderness for hiking, camping, and nature study.[18]

Women moved into the void. They created their own groups, promoted their own agendas, and crafted their own strategies, which in many cases differed significantly from their male counterparts. Established conservation groups tended toward structure and formal organization. Women's organizations often took a grassroots, bottom-up approach. In doing so they facilitated a new kind of citizen activism focused on the environment, and they played a pivotal—and underappreciated—role in fueling modern environmentalism and ultimately a global movement for ecological change. It took some time for establishment conservationists to recognize how thoroughly women were reshaping the political narrative on the environment. As Glenda Riley has noted, the women's initial tendency to move cautiously away from domesticity and toward activism made it easy to overlook them, at least initially.[19]

The most prominent and significant, undeniably, was Rachel Carson, the scientist and writer whose 1962 book, *Silent Spring*, ignited a national dialogue about the dangers of pesticides, particularly dichlorodiphenyltrichloroethane (DDT). Initially used during World War II to kill mosquitos that carried malaria, after the war DDT was sold for domestic use, to kill weeds and control pests, including gnats and certain kinds of fish that overpopulated waterways. Carson began her research on what would become *Silent Spring* in the late 1950s.[20] In speeches, articles, and testimony to Congress, she described a nation awash in "an appalling deluge of chemical pollution." And these were not "selective poisons.... They do not single out the one species of which we desire to be rid." Instead they are "passed from one organism to another" through the food chain, found even in breast milk. "Only within the moment of time represented by the present century has one species—man—acquired significant power to alter the nature of his world."[21]

Carson's earlier books about the sea had become best-sellers that drew plaudits and uniformly positive reviews. However, *Silent Spring* challenged the industrial-corporate complex; consequently she faced immediate and strong blowback from powerful interests. She was accused of being both overly emotional and a communist. As an unmarried woman, she could be framed as challenging male authority, "drawing on a long legacy of gender inequity." But the book emboldened some women to speak out, such as one in San Bernardino, California, who "argued that 'pesticide manufacturers making billions from their poisons should be ... required to prove that there are no dangers from eating mixtures of poison residues at every meal.'"[22] Ultimately the book planted the seeds of what soon would become the environmental movement. And it put government and industry on notice that ordinary people had powerful voices in policies that impacted their health and well-being.[23]

There were other women as well. In Florida, Marjory Stoneman Douglas worked to save the Everglades from development. A newspaper and magazine writer in Miami for many years, in 1947 she published *River of Grass*, a book about the Everglades, soon to be dedicated as a national park. She had known little about the area, she admitted, except that it was a place with many unique plants and birds. In the 1960s Douglas learned of a plan to build an airport on landfill at the edge of the park and became actively involved in the effort to stop it. As she explained in her autobiography, *Voice of the River,* "Man has been doing his best to drain, plug, stanch, dike, and otherwise remove the water from the Everglades since the beginning of this century. . . . The engineers drained to satisfy the sugar people, then drained some more to satisfy the cattle people, who wanted more dry land to support the dairy herds." She added, "The dairy herds produced manure, and the manure found its way into the rivers."[24] In her midseventies she traveled throughout the state, giving speeches. She penned editorials and lobbied public officials. The jetport was abandoned and restoration work inaugurated.[25]

Jane Jacobs was the married mother of three children in the early 1960s when she took on the New York City establishment over plans to raze neighborhoods in Greenwich Village to build an expressway and create urban renewal projects. In the process she became a vociferous and effective critic of Robert Moses, the powerful master builder of New York. Jacobs did not lean on motherhood to make her case, though Moses derisively depicted her in those terms. At one point he railed, "There's nobody against this [project]—nobody, nobody, nobody but a bunch of mothers." Jacobs's 1961 book, *The Death and Life of Great American Cities,* placed environmental concerns in relatable terms.[26]

In Alaska, Margaret "Mardy" Murie worked alongside her husband, Olaus, and others to preserve what became the Arctic National Wildlife Refuge. "It began to appear that even the vastness of Alaska's wilderness would not remain unexploited without some special legal protection," she wrote in 1956. "The Age of the Bulldozer had arrived."[27] And in Arizona a folksinger named Katie Lee wrote songs about the destruction of Glen Canyon due to the construction of a 710-foot high dam. "When they drowned that place, they drowned my whole guts," she said in an interview. At one point Lee released an album that "pilloried the Bureau of Reclamation."[28]

Carson, Douglas, Jacobs, Murie, and Lee largely came to their environmental work as individuals, but many postwar California women became involved through established community connections. Some belonged to national organizations that reflected domestic concerns, such as Women Strike for

Peace. WSP lobbied against atmospheric nuclear testing, specifically targeting strontium-90, a byproduct of testing that drifted to earth, where winds spread it to farmlands and fields. Livestock ate the grasses, so strontium-90 ended up in food and milk. WSP had several California chapters. Members pushed babies in strollers as they held aloft signs during protests. A number of WSP members sent their children's baby teeth, infused with radioactive particles, to lawmakers, hoping to convince them to stop nuclear testing.

La Wisp, the Los Angeles chapter newsletter, advised members to dress modestly when gathered in public. "We made a distinctly good impression on all those who saw us walking, and received several fine compliments," a WSP writer exulted after a protest. "One of the reporters who interviewed us went out of his way to tell us that back at the Press Club the guys had decided that we were the prettiest picket line they had ever seen." *La Wisp* even published a cookbook, *Peace de Resistance*. One member, Amy Swerdlow, acknowledged that the maternal approach might seem "superficial and frivolous" to many, but at a time when anticommunist rhetoric still held the power to instill fear, it was "a highly political statement . . . [that sought] to assure those who were fearful of Red baiting or social disapproval that WSP could be both militant and lovable." Concerns about Red-baiting were in fact justified. Called to testify before the House Committee on Un-American Activities, WSP members declared themselves to be "apolitical mothers who sought only to protect their children from radiation."[29]

Postwar women shaped a new kind of environmental movement. As in many other endeavors, however, the work of earlier generations had laid the foundation. From its earliest days of statehood, California attracted a number of remarkable women—writers, scientists, climbers, and reformers among them—who passionately embraced nature and its offerings. Mary Hunter arrived in California from Illinois in the late 1880s with her widowed mother and siblings. The family homesteaded near the southern San Joaquin Valley town of Bakersfield. She was twenty when she met Stafford Wallace Austin, whom she soon would marry. The Austins left Bakersfield for the San Francisco Bay Area, and then settled successively in several towns in Owens Valley, in eastern California. The valley sat between two mountain ranges, the Sierras and the White Mountains. Both of the Austins became educators, and Mary began to write about her new home. "The West for Austin . . . shaped the ways she thought about America and about herself," wrote her biographers. She reflected the emergence "of California and the Southwest in the national conscience."[30]

Austin wrote the first of her thirty books in 1900 and published her best-known book, *Land of Little Rain*, three years later. To understand "the fashion of any life," she wrote, "one must know the land it is lived in and the procession of the year."[31] Speaking of the valley, she wrote, "None other than this long brown land lays such a hold on the affections. The rainbow hills, the tender bluish mists, the luminous radiance of the spring, have the lotus charms. They trick the sense of time, so that once inhabiting here you always mean to go away without quite realizing that you have not done it." She also wrote about Native Americans. One, called "Walking Woman," roamed "the southern San Joaquin" Valley.[32] And she wrote about animals, comparing their behavior—favorably—to that of humans: "Man is a great blunderer going about in the woods. . . . Being so well warned beforehand, it is a very stupid animal, or a very bold one, that cannot keep safely hid. The cunningest hunter is hunted in turn, and what he leaves of his kill is meat for some other. That is the economy of nature. . . . There is no scavenger that eats tin cans, and no wild thing leaves a like disfigurement on the forest floor."[33]

In 1905 the Austins became embroiled in the Owens Valley water wars, which began after Los Angeles officials, concealing their motives and their identities, convinced valley farmers to sell them land abutting the lush Owens River. Long a sleepy Mexican and then American town, Los Angeles began to transform into a city with the arrival of the railroad in the 1870s. By the turn of the twentieth century, it had more than 100,000 residents. Since the region was subject to bouts of heavy rain followed by prolonged droughts, it became increasingly clear that any future growth would be limited without access to a steady water supply from somewhere else. And Los Angeles leaders definitely wanted growth.

Owens Valley, nearly three hundred miles from Los Angeles, was that somewhere else. When Los Angeles officials had purchased enough farm land, they announced their plan to build a pipeline from Owens Valley to Southern California and sought a bond measure for that purpose. Dire warnings about drought and ruin propelled voter approval, and the 233-mile pipeline was completed in 1913. Austin, along with many others, decried what they called "the theft" of the water. She wrote articles in the *San Francisco Chronicle* decrying "the chicanery" of Los Angeles, and later called it a "very wicked episode."[34] By 1920 the population of Los Angeles had grown to nearly 600,000. Meanwhile Owens Valley largely turned into a desert.

After leaving her husband, Austin moved to San Francisco, just in time for the earthquake that leveled much of the city. She then settled for a time in the seaside artist community of Carmel, where she became part of a group of writers that included Robinson Jeffers, Lincoln Steffens and his wife Ella Winter, and Jack London. Once again she painted landscape portraits with words. Carmel, for example, was a "virgin thicket of buckhorn sage and sea blue lilac between well-spaced, long-leaved pines." Austin eventually left California and spent the last few years of her life in Santa Fe, New Mexico.[35]

Female scientists also arrived early in California. They included Mary Katharine Brandegee, whose family moved to the new state during the Gold Rush. In the 1870s she became the third woman to graduate from the University of California Medical School, but she chose to become a botanist instead of a doctor. Botany, as Susan Schrepfer argues, gave women an entree into science, since studying plants was viewed as a less masculine and thus more socially acceptable pursuit than geology or biology. Brandegee wrote extensively on California plants and curated the collection for the Academy of Sciences in San Francisco.[36] Annie Alexander moved to Oakland with her family in her teens. She attended college in Massachusetts, then returned to California. After hearing a lecture on paleontology at the University of California, Berkeley, she decided to devote her life to the study of fossils. Using family money, she underwrote the Museum of Paleontology at Berkeley and took part in many expeditions. Following one trip she wrote, "We worked hard up to the last, marking and wrapping bones." She and another female paleontologist cooked for their male colleagues on the trips: "Night after night we stood before a hot fire to stir rice, or beans, or corn, or soup . . . We sometimes wondered if the men thought the fire wood dropped out of the sky."[37]

Kate Sessions graduated from UC Berkeley in 1881. Her graduation essay was titled "The Natural Sciences as a Field for Women's Labor." After moving to San Diego, she helped to create the San Diego Floral Society. She leased thirty acres of land and planted hundreds of trees each year and came to be called "the mother of Balboa Park."[38] Ellen Browning Scripps also settled in the San Diego area. A scion of the Scripps newspaper family, she moved to California from the Midwest in the 1890s. A strong advocate of women's rights and scientific exploration, in 1903 she financed the San Diego Marine Biological Association, designed to "carry on a biological and hydrographic survey of the waters of the Pacific Ocean adjacent to the coast of Southern California ... [and] to build and maintain a public aquarium and museum." When she moved to nearby La Jolla, that town became the site of the Scripps Institution for Biological Research, affiliated with the University of California. Scripps also financed Torrey Pines State Reserve, a two-thousand-acre park winding through windswept bluffs and coastal canyons, as well as a lodge at Asilomar on the Monterey Peninsula. The architect Julia Morgan built the structures.[39]

Alice Eastwood was a self-taught botanist. Born in Canada, she taught high school in Denver before moving to San Francisco in 1891. A year later she became curator of the California Academy of Sciences. In that capacity she spent large chunks of time exploring the West, often alone, searching out specimens for the academy. Wearing bulky women's clothing made the task more difficult, so she fashioned a skirt with buttons linking front to back, creating an early form of culottes. When an April 1906 earthquake leveled most of San Francisco, including the building that housed the science academy, Eastwood, climbed the still-standing bronze bannister hand over hand to rescue specimens; altogether she saved nearly 1,500 of them. The earthquake "did not frighten [her]," she later said, but it did cause a "great loss to the scientific world and an irreparable loss to California."[40] During her long life, she authored more than three hundred research papers and a guidebook to wildflowers. As the most prominent female botanist in the West, she also helped to plan San Francisco's Golden Gate Park.[41]

Women also joined the Sierra Club in the early twentieth century, participating in arduous backcountry trips that often lasted as long as a month. Nora Evans had been teaching math in Visalia, California, when she made her first trip into the high country near Yosemite in 1911. "In those days there were no roads," she explained, "and you were lucky if you got up to 3,000 feet from the west side." She described wearing bloomers and middy blouses—women did not wear pants on hikes until the 1920s—and making her own sleeping bags out of eiderdown or wool blankets. Hikers generally slept out of doors. At lower elevations the trees "were heavy enough that you could keep pretty dry." Once, she "foolishly kept some bacon by [her] pillow": "I woke up at dawn and there was a huge bear about fifty feet away." During her long hiking career, Evans climbed Mt. Whitney in California and Mt. Rainier in Washington. On one climb she was struck by lightning: "[Afterward] my feet felt as though I was walking on hot coals."[42]

Elizabeth Marston Bade graduated from Wellesley College and moved to San Diego. She made her first high-country trip in 1906, to Kings River Canyon, taking a train from San Diego to Visalia, then riding a stagecoach into the mountains. "We left early in the morning in a four-horse stage and drove over the mountains to [the town] of Millwood," she wrote. "We started off walking into the west bank of the King's River. It was flooded, so we had to sit on the bank waiting for the rangers to fix the bridge." On another trip, hikers encountered many rattlesnakes. "I had been majoring in zoology in college," Bade recalled, "so with great joy I helped . . . cut up a rattlesnake which had swallowed a gopher." She continued high-country trips after meeting and marrying her husband, William Bade, who served as executor of John Muir's papers and authored the book *Life and Letters of John Muir*. At one point the Bades decided to replicate Muir's journey across the United States to California, traveling to the East Coast, to Florida, Cuba, and then back across the country.[43]

The Progressive movement of the early twentieth century provided opportunities for women to work in politics as reformers, both nationally and locally. Progressives sought to reorder society, tackling political corruption, urban decay, and social disintegration through new institutions and agencies. Preserving and expanding natural resources played a prominent role in this ambitious agenda. It was "a defining feature of progressivism—and particularly important for the arid West."[44] Conservation fit squarely into women's agenda. White men who built businesses and ran factories were complicit in creating many of the issues Progressivism aimed to solve. Women worked to ameliorate such problems by establishing settlement houses and calling for child labor laws, food and drug legislation, expanded educational opportunities, temperance, and suffrage. In California, Progressive women also fought for prison reform, juvenile courts, and public education through high school.[45]

"Nature preservation . . . was a major factor that aided women's mobilization," asserts Cameron Binkley.[46] Carolyn Merchant suggests, "Leisure time . . . afforded middle and upper class women opportunities for botanizing, gardening, bird lore, and camping. Propelled by a growing consciousness of the panacea of bucolic scenery and wilderness, coupled with the need for reform of the squalor of the cities, women burst vividly into the public arena in the early twentieth century as a force in the progressive conservation crusade." Often they used women's clubs as a vehicle for reform.[47]

Laura Lyon White—Mrs. Lovell White, in the parlance of the day—participated in several campaigns. Born in Indiana and reared in Iowa, White came to California in the 1850s with her husband. The couple lived in a mining town for a time, then relocated to San Francisco, where Lovell White became a wealthy financier. By the late 1800s Laura White was an ardent suffragist; after Californians defeated women's suffrage in the 1890s, she focused her prodigious energy on conserving natural resources. In 1900, with Clara Bradley Burdette and others, she founded the California Federation of Women's Clubs (CFWC). Burdette lived in Pasadena. She once declared that women "must go beyond the home": "No longer is the home encompassed by four walls. Many of its important activities lie now involved in the bigger family of the city and the state."[48]

The group's first campaign centered on the Calaveras Big Trees, old-growth sequoias in the Sierras northeast of the city of Stockton. After a Minnesota lumberman bought the grove in 1900, White initiated a campaign to convince Congress to purchase the grove for a national park. She traveled to Washington DC to lobby before natural resource committees. Congress favored the proposal, but the owner refused to sell. In 1904 White launched a national petition drive that garnered 1.4 million signatures and led President Theodore Roosevelt to declare the trees a "national inheritance." The owner still refused to sell, but left the trees standing. The women continued lobbying, and in 1909 Congress passed a bill, signed by Roosevelt, authorizing the exchange of the sequoias for federally owned lands elsewhere. It took another two decades for the arrangement to be finalized.[49]

Working with both women and men, White and the CFWC also participated in a campaign to stop loggers from destroying old growth redwoods in the Big Basin area of the Santa Cruz Mountains. Women did grassroots organizing, while men leaned on their political connections. The group decided to seek state park status so as not to interfere with the national Calaveras effort. Andrew Hill, a San Jose photographer, and Josephine McCracken, a Santa Cruz writer, initiated the effort. Hill approached women's clubs, specifically the San Jose Women's Club, for help. Newspaper editors, educators, the CFWC, and the region's Pioneer Society got involved. So did Phoebe Hearst, mother of the publishing titan William Randolph Hearst, who helped to bankroll the campaign.

Women's club members penned editorials and letters to the editor decrying the "wastefulness" of logging. They circulated petitions and formed the Sempervirens Club, devoted to preserving redwoods. In 1901 California's governor Henry Gage signed legislation authorizing the park, but state lawmakers continued to allow logging. The area finally achieved state park status after the Progressive Hiram Johnson became governor in 1911. Seven years later the Sempervirens Club became the Save the Redwoods League, headquartered in California.[50]

Women's club members also worked with Muir and the Sierra Club to stop the damming and flooding of Hetch Hetchy to provide water to San Francisco. Beginning in 1909, after San Franciscans approved a referendum in favor of a dam, they wrote letters and sent telegrams to members of the U.S. House of Representatives Committee on Public Lands. "Among them were women who had camped and hiked in the valley, who were members of the Sierra Club or the Appalachian Mountain Club, and who were opposed to the commercial use of such a scenic wonderland." The effort failed in 1913, when Congress authorized the dam.[51]

After World War I more women participated in conservation activities. Advertisers began to feature women climbers in publications; no longer clad in skirts or bloomers, they now wore knickers, "wide at the hip and fitted at the knees."[52] And at a time when women made up a relatively small percentage of Sierra Club membership, a few assumed leadership positions. Aurelia Harwood came to California from Missouri. Never married and in her fifties, she lived with and cared for her wealthy elderly parents; her father had been president of the Ontario-Cucamonga Fruit Exchange. In the early part of the twentieth century, Harwood worked to save the redwoods from logging and lobbied California lawmakers to ban hunting in national parks, and she purchased mountain land and donated it to the National Park Service. An avid climber and leader of the Campfire Girls, she led hikes into the mountains of Southern California. In 1927 she was elected the first female president of the Sierra Club. She died suddenly during her second term in 1928. In her honor the club built a lodge in the Angeles National Forest at the base of Mt. Baldy.[53]

Marjory Farquhar was considered one of the world's foremost female climbers; she scaled all of the fourteen-thousand-foot-high peaks in the West except Mt. Rainier. Born in 1903 and reared in the Marin County town of Mill Valley, she learned to climb early, by shimmying up the mast of her family's sailboat. On many weekends her father led sailing trips to the Sacramento River, about ninety miles to the east. "The river was such that we had these levees and from a boat . . . you couldn't see over the levees to see what was growing," she said. So she scaled the mast for a better view.[54]

At the University of California, Berkeley, Farquhar studied social work. After graduation, she traveled to Europe with a group from the YWCA. She began taking photos and, when she returned to the United States, embarked on a career as a photographer, mostly of children and of weddings. At the time, photography largely consisted of portraiture, she recalled, but that didn't satisfy her: "I wanted pictures where [subjects] were doing things." She also began to get serious about climbing mountains. For her first pack trip in 1929, she took a train from San Francisco to Fresno, then took a bus to Florence Lake near an area of the Sierras called Evolution Valley. "Mountain climbing was so wonderful . . . such an exhilarating sport."[55]

In 1930 she began climbing in earnest. Goddard Mountain, about fifteen miles east of Florence Lake, was nearly fourteen thousand feet high. She and a group of female hikers "knapsacked up out of Colby Meadow and stayed all night. Then [they] went over and climbed the north face of Goddard." The following year she began climbing with ropes, and in 1934 she ascended the east face of Mt. Whitney, a feat she repeated several more times. On her third climb she carried a movie camera. Farquhar married that same year; her husband, Francis, was a world-renowned climber.[56]

Farquhar also skied in the backcountry, a difficult task with few ski lifts, lodges, or passable roads. "Once we slept in the old station at Soda Springs," she recalled. At one point the Sierra Club skiers decided to build a rustic lodge. The women may have been partners in mountaineering, but traditional gender roles prevailed in other aspects: men built the lodge, while the "girls would cook and provide all the meals for the laborers." Farquhar stopped working as a photographer and climbing regularly in the 1940s. "I realized I couldn't do it all," she told an interviewer. In the early 1950s she was elected to the Sierra Club board of directors, one of very few women to achieve this honor. Asked later about her experiences as a woman in the Sierra Club, she replied, "I just felt I was as good as anybody else." Farquhar professed the belief that spending time in the outdoors built "dedicated conservationists." If people "knew what they were talking about, they could fight for it."[57]

As Farquhar's climbing career wound down, women in the San Francisco Bay area joined forces to preserve open spaces and to lobby for controlled building and development. Their efforts signaled a new and broader agenda on behalf of the environment; rather than focusing entirely on rural areas, they sought to preserve open spaces in urban or near-urban areas. The completion of two monumental Bay Area projects—the Bay Bridge, linking San Francisco and the East Bay, and the Golden Gate bridge, connecting San Francisco and Marin County to the north—served as catalysts for the shift. Ferries had previously been the only means of transportation across the bay in both directions. The East Bay, led by Oakland, was already developed, but Marin County was largely isolated and rural, dotted with small, upscale towns mostly on the eastern side. The western side featured bays, pristine beaches, towering cliffs, and mountains offering spectacular views.

In 1934 four prominent Marin County women, concerned that the planned Golden Gate Bridge would "bring an influx [of traffic] that would jeopardize the county's open hills and valleys," created the Marin Conservation League.[58] Caroline Sealy Livermore served as president of the Conservation League for twenty years. Originally from Texas, she was married to Norman Livermore and the mother of five sons. The family lived in the Marin County town of Ross. She was "very intelligent and sharp. . . . She grew up before women were generally accepted in business, but she was one of those women who made her own place in the world," writes Richard A. Walker. As another Bay Area woman activist said admiringly, "Caroline was a master."[59]

Livermore and her colleagues began by raising $2,500 to pay for a study to guide county officials in planning for growth. They prodded officials to adopt a master plan on development, using federal funds from President Franklin Roosevelt's New Deal. They lobbied supervisors on the need to hire a permanent county planning director, and in 1941 the county hired Mary Summers to head the planning department. Summers had earned a landscape architecture degree from the University of Oregon and initially worked as a secretary in the planning department of Contra Costa County. She got the Marin County job, she recalled much later, because her male predecessor left to join the military. In an oral history, she described difficulties initially getting men to take her seriously: "No engineer wanted to talk to me. I was just a kid. I was 23 years old. How could I know anything?" Summers stayed in the job for twenty years and became one of Livermore's closest allies.[60]

The Conservation League campaigned, successfully, for land to be set aside for state parks throughout the region. By the 1940s the League had helped to create parks in some of Marin County's—and California's—most beautiful areas: Mt. Tamalpais, Stinson Beach, Tomales Bay, the Marin Headlands, Point Reyes. "Every highway, bay-fill, and army-base project triggered a response from the Warriors of the Marin Conservation League and a burgeoning group of allies," which came to include the Sierra Club, the Audubon Society, and the Nature Conservancy, notes one writer.[61]

One of Livermore's campaigns aimed to save Angel Island from developers. Situated in the bay between the city of San Francisco and Marin County, Angel Island had been a cattle ranch and a military base. From 1910 to 1940 it served as a detention and processing center for more than a million immigrants hoping to gain entry to the United States via San Francisco.

The Chinese constituted the largest group, but in its thirty-year history the island housed immigrants from eighty countries. They were taken from ships in San Francisco Bay, processed, and then sent to an island hospital to be examined by doctors who required them to defecate into washbasins. Their belongings were confiscated, and they were assigned to dilapidated and dirty barracks. Detainees also suffered from poor ventilation and a lack of hot water. Many were kept for months. If they were deemed medically unfit—pregnancy, hernias, or heart disease qualified—they could be sent back to their country of origin. During World War II, Angel Island housed prisoners of war. Afterward it was abandoned. Its prime location made it an enticing target for developers. In the early 1960s Livermore formed the Angel Island Foundation to convince the federal government to sell it to California as a state park; it won that designation in 1963. The island's highest peak is named Mt. Caroline Livermore.[62]

Many people in Marin County cheered the work of Livermore and the Conservation League, but some critics argued that the well-off white women strove to keep pristine areas for themselves and to keep everyone else out. This refrain would continue to dog environmentalists, both women and men, through the coming years. Grace Wellman, a League member, acknowledged this complaint: "They said we'd like to lock the bridge and keep people out." Not surprisingly, she disagreed. The League "didn't try to keep people out, [it] tried to plan for them so that when they came in, there'd be a . . . way for these people to live."[63]

While Livermore pressed for structured planning and open space in Marin County, Dorothy Ward Erskine was engaged in a similar enterprise across the bay in San Francisco, a city that had shed its rural roots long before. The population of Marin County in the 1940s was slightly more than 50,000; San Francisco City and County held more than 600,000 residents.[64] Whites were a majority in both areas, but San Francisco had a large working-class population. Thus Erskine's work can be viewed in terms somewhat different from Livermore's. The latter worked to save open space in a place that had not yet lost it, while the former worked to save what little remained.

Erskine's background also differed significantly from that of Livermore, though the two women were good friends. She came from a Quaker family and her mother, Florence, was a homeopathic physician who delivered three of Livermore's sons.[65] She attended the University of California, Berkeley, and married a lawyer. In the 1930s she flirted with communism, traveling twice to the Soviet Union, before becoming disenchanted. She was friendly with a group of bohemian writers and artists who lived in Los Gatos and in Carmel, and she was a devotee of Alice Griffith, who, following Jane Addams's Chicago model, created a settlement house in San Francisco in the 1890s. After the devastating 1906 earthquake, Griffith was instrumental in creating the San Francisco Housing Association, aimed at passing legislation "to prevent jerry-building in the reconstruction of the city."[66]

Erskine's involvement in city and county politics started in the 1930s. "It began with slums," she told an interviewer, after Congress had passed public housing legislation. "We formed a kind of economic study group." Participants successfully lobbied the California State Legislature to provide federal subsidies for low-income housing in the city. Aided by a group of UC Berkeley graduates, Erskine and her allies began to focus on urban planning as well. The students "put together an exhibit" featured at the San Francisco Museum. "You say city planning: well, you don't begin with city planning—at least we didn't here—until the city is almost completely built up," Erskine said. Within a few years the group had created the San Francisco Planning and Urban Renewal Association.[67]

Erskine also worked to preserve green open spaces in San Francisco and the larger Bay Area. In the 1950s she founded Citizens for Regional Recreation and Parks (CRRP), an effort that drew in the Sierra Club, the Marin Conservation League, and others. When the group learned that the East Bay Municipal Utility District planned to build houses on 6,500 acres surrounding the San Pablo Dam near the town of El Sobrante, CRRP successfully fought the proposal. Ultimately CRRP morphed into People for Open Space, and then into the Greenbelt Alliance. Early on, Erskine admitted, they "were always baffled" by the fact that other cities had access to large amounts of money for planning, while CRRP did not. On one trip to the East Coast with her husband, Morse, she decided to take the train while he flew home. Stopping in major cities along the way, she learned how other activist groups raised money; in short, they asked businesses, foundations, and government entities for donations. When she returned home, she too began asking for money; at one point she snagged a grant from the Ford Foundation. Despite all of her accomplishments, Erskine was "typical of women of her time," writes Walker, "self-effacing and happy to shift credit onto others." Today Dorothy W. Erskine Park sits atop a hill in San Francisco's Mission District.[68]

By the 1950s a critical mass was forming on behalf of conservation across the country as well as in California. As the *New York Times* reported, "People generally—and that means voters—are gradually awakening to the fact that the natural resources of their country, and much of its natural beauty, have been disappearing before their eyes. . . . There is also much evidence of a growing determination to do something about the problem." The *Times* story took note of several campaigns, highlighting the successful effort by the Sierra Club to stop construction of a dam in Dinosaur National Monument. But there were many other issues as well. In Indiana "each day, US Steel pumped 300 million gallons of waste water into Lake Michigan and the Grand Calumet River." In New York City strong opposition had emerged to a plan to fence in a "picturesque area" in Central Park known widely as "an extraordinary urban observation post for the flight of migratory birds."[69]

The growing emphasis on environmental issues "represented an unprecedented attempt to reclaim the nation's natural resources," argues Andrew Hurley. Throughout the country, as groups and individuals homed in on a wide variety of issues, women began moving to the forefront of the movement. In the Gary, Indiana, suburban neighborhood of Miller, for example, Helen Hoock, Naomi Stern, and others labored to stop a local utility from building a 500-megawatt power plant, which would spew 105,000 tons of sulfur dioxide into the air each year.[70]

As the fastest growing state, California had to contend with a wider array of issues than most others: smog, sprawl, traffic gridlock, fires, landslides, water pollution, pesticides, threats to the ocean from oil drilling, fears about the safety of proposed nuclear power plants, diminishing amounts of open space. As a result, women began taking on crucial and highly visible roles in the movement that was morphing from conservation into environmentalism and moving from the margins of political discourse to the center.

The timing was right for such a shift. During World War II women had been called into service in many jobs previously held by men: building and flying airplanes, building ships, playing professional baseball, performing in "all-girl" orchestras and in rodeos. War workers, including thousands of women, eventually found their way to California. Many lost their jobs when the soldiers came home. Some rejoined the labor force within a few years; others pursued volunteer opportunities in their community.

Potential volunteers did not lack for causes, though many, like Sylvia McLaughlin of Save the San Francisco Bay, had no idea how committed they eventually would become or how their involvement might change their lives and that of their community. Volunteering seemed a natural fit. Women could sell themselves to the general public, and to public officials, as working for the common good rather than, as the songwriter Katie Lee put it when talking about the proposal to dam Glen Canyon, working just for "me-me-me." Of course, they ran into obstacles, including from corporate titans who resented being called to task—by housewives, no less—for polluting the environment, and from politicians who profited from corporate donations. Over time the activists learned how to adapt their strategies to political realities, and also how to use their hard-won knowledge to change politics.

Some even became part of the political power structure. Jacqueline Rynerson moved to Lakewood with her family in 1952 and became involved in a campaign to ensure that the much larger city of Long Beach did not swallow up its smaller neighbor. Soon she was working on city beautification projects and campaigns to create parks, swimming pools, and hiking trails. In the 1970s she ran for and won election to the Lakewood City Council, where she served three terms. She also served as the city's mayor and represented Lakewood on the Southern California Association of Governments. Rynerson's daughter, Julie, later recalled, "Some of my earliest memories were of sleeping in the back of the station wagon while they would have these neighborhood meetings to organize people to go out and petition." Today Rynerson Park encompasses forty acres in northern Lakewood. It includes trails, wildflowers, playgrounds, and baseball fields.[71]

In an era that extolled stay-at-home wives and mothers, the activism of Rynerson and others might have been expected to draw criticism, but, as usual, reality was much more complex. Magazines, newspapers, and television programs may generally have depicted domesticity as the ideal lifestyle, but opinion polls at the time revealed a different narrative. Asked to name women they admired, respondents most commonly cited the former first lady Eleanor Roosevelt; Helen Keller, a writer and advocate for people with disabilities; and the author Clare Booth Luce, all of whom reflected "courage, spirit, and conviction." And readers applauded women's participation in politics, declaring, "Voting, office-holding, raising your voice for new and better laws are just as important to your home and family as the evening meal or spring house cleaning."[72] Many women in the environmental movement reflected both aspects: they were wives and mothers, but also activists. By the early 1960s they were working on numerous grassroots environmental campaigns all over California, in the process nudging male leaders of long-standing mainstream organizations and local and state politicians to broaden or rethink their agendas.

Jean Kortum was the wife of a Democratic Party activist and the mother of three, living in San Francisco. Originally from Iowa, she graduated from Pomona College in Southern California and then moved north to the San Francisco Bay area, where she took a reporting job at an Oakland newspaper. She became active in Democratic Party politics, but mostly behind the scenes. As one writer puts it, "Expected to fill the role of middle-class housewives, women such as Kortum nevertheless pursued outlets for their talents."[73] In 1959 she became involved in the campaign to stop PG&E from building a nuclear power plant in Bodega Bay, a scenic area about fifty miles north of San Francisco. Working against the area's business and political establishments, Kortum and others created the Northern California Association to preserve Bodega Head and Harbor. She recruited people to picket PG&E headquarters in San Francisco, carrying signs bearing messages such as "Iodine and Milk Don't Mix." At one point the group inflated and released hundreds of balloons bearing the message "If it had been strontium 90, it would have landed in the same place."[74]

Kortum lobbied Bay Area officials to oppose the power plant. When they refused to work with her, she called them "gutless." She soon realized that the California Democratic Party could be a powerful ally. After an intense lobbying and letter-writing campaign, she won support from the California Democratic Council, a statewide organization comprising Democratic clubs across the state. She lobbied Lieutenant Governor Glenn Anderson, a Democrat, to write to the Atomic Energy Commission, urging it to withhold approval of PG&E's plans for the power plant. Kortum also convinced Phillip Burton, a powerful California Democratic state assembly member—later a congressman and close ally—to sponsor legislation and to lobby Interior Secretary Stewart Udall to oppose the plant. The constant pressure began to work, and by the early 1960s public criticism over the project had grown significantly. Protests at the proposed site drew hundreds, and some politicians once committed to the project began to back down. In the end, the discovery of an earthquake fault doomed the plant, but the work of Kortum and others set the stage for PG&E to abandon the project.

Kortum also played a prominent role in an effort to stop plans to build a number of freeways through San Francisco. In the late 1940s the California Division of Highways drew up plans for a web of freeways around the Bay Area. In response, the San Francisco Planning Commission "called for 25 miles of 8 elevated skyways" throughout the city. At the time, freeways "reigned supreme," writes Chris Carlsson, "but San Francisco harbored the seeds of an incipient revolt ... the first serious opposition to the post World War II consensus on automobiles, freeways and suburbanization." The fight to stop freeway building came to be known as the "Freeway Revolt."[75]

In 1956 the *San Francisco Chronicle* published a map showing the proposed freeways, which would encircle and bisect the city. Local opposition soon emerged, with Kortum as a leader of the resistance. Using "the telephone to rally her troops and deluge political leaders with calls and letters," Kortum organized the Freeway Crisis Committee, which brought together neighborhood groups across the city under the slogan "Save Our City." Picketers marched through the streets and showed up at supervisors' meetings bearing signs and wearing paper hats that read "Save Us from Freeways." They submitted petitions containing thirty thousand names.. Under pressure, supervisors dropped their support for seven of the proposed freeways, but the dance among city-county, state, and federal officials continued.

Some freeway supporters charged that opponents were merely upper-middle-class homeowners "concerned with the view," but the proposed freeways would impact low-income areas as well, including one that would cut through neighborhoods adjacent to Golden Gate Park. A massive May 1964 rally in the park marked a turning point, though it would take nearly two more years to definitively defeat the freeway proposals. The folksinger Malvina Reynolds attended the rally. A song she wrote for the occasion reflected the residents' anger and frustration; called "Cement Octopus," it referenced the state's politicians.[76]

Governor Edmund G. "Pat" Brown was a strong advocate of growth and expressed disappointment with the initial outcome, calling the freeway failure a tragedy. A *San Francisco Chronicle* columnist later called Kortum "an unsung hero of the Bay Area environmental movement . . . indefatigable," part of "a generation . . . who revitalized the conservation movement in the 1960s and 1970s."[77] Ultimately the San Francisco "Freeway Revolt" led to other, similar efforts in places like Boston, and it helped to foster the creation of the public transportation system, Bay Area Rapid Transit.[78]

Meanwhile, across the bay to the east, in the largely industrial and working-class city of Richmond, Lucretia Edwards worked to increase the amount of coastline access available to the public. Edwards moved to Richmond in the late 1940s with her husband, Thomas, a docking pilot for Standard Oil Company. She was stunned to discover that, with miles of coastline, less than one hundred feet was available for public use. Several major corporations, including Standard Oil, Kaiser Shipyards, and Union Pacific Railroad, owned the rest. "You hardly knew the bay was there," she told one interviewer.[79]

Edwards joined the League of Women Voters and set to work, lobbying for increased public access via shoreline parks. She spent "endless hours" at the offices of city, county, and state officials, talking to "any officials who would listen." At some point, she and her League colleagues struck on a novel idea: taking officials to the "shoreline" for picnics, then engaging in a "soft-sell" effort. As she recalled much later, "We very much believed in the divide-and-conquer theory. We got dressed up in our flowery hats and invited local politicians and officials to picnics with lots of cheap champagne. Once

39

they were comfortable, we would begin the process of convincing them of the value of public access."[80]

Success came slowly. At one point a landowner reneged on what Edwards believed had been a promise to sell his land for a park. Instead he decided to sell to a developer. She was so distraught, her husband bought the land and then sold it to the Shoreline Park Committee. Today parks line the shore. They include the Miller/Knox Regional Shoreline, a park of nearly three hundred acres; Point Pinole Regional Park, with more than 2,300 acres; and the Rosie the Riveter/World War II Home Front National Historical Park, dedicated to women defense workers. Richmond had been a center of ship-building during the war, and many of the workers there were women. At a 2004 ceremony honoring her, Edwards credited her Quaker background for her involvement in the world outside her home. She added, "I was expected to be polite, lady like and not to be a nuisance to anybody. But I soon learned what great fun it is to antagonize people."[81]

By the 1970s the environmental movement had put down deep roots, with women leading numerous campaigns throughout California. Dorothy Green moved from Michigan in the 1940s to attend UCLA, where she majored in music. After graduating in 1951, she married, settled in Westwood, and reared three children. Throughout the 1960s she was involved in a number of civil rights groups. She first became active in environmental causes in 1972, with the successful campaign to pass Proposition 20, the measure creating the California Coastal Commission. "It really was a special time that deserves to be remembered," she said. "I was fortunate enough to have become active just as the decade started, because it was that whole decade ... where volunteers went out, gathered signatures, and then ran campaigns."[82]

In the early 1980s Green worked to defeat the Peripheral Canal, a California ballot initiative that proposed to divert water from the Sacramento River Delta to the Central Valley and Southern California. Opponents argued that the measure would subsidize so-called Big Ag farmers at the expense of Delta residents and smaller farmers. It failed by a whopping 63–37 percent. Her most significant campaign began in the mid-1980s, when she learned that millions of gallons of barely treated raw sewage were being dumped into Ballona Creek, which drains into Santa Monica Bay. Swimmers in the bay—spanning more than four hundred square miles from Malibu to the Palos Verdes Peninsula—complained of foul odors, skin rashes, and ear infections and described seeing dead fish with large tumors. Since the 1920s Hyperion Sewage Treatment Plant had contracted with the city of Los Angeles to handle the sewage, and clean-up efforts had been made, but the toxic sludge—from power plants and Los Angeles International Airport, among other entities—continued to contaminate the water. In 1985 environmentalists protested Los Angeles's request to be exempted from parts of the federal Clean Water Act. Green gathered a small group of environmentalists in her home and plotted strategy to oppose the city's request. "I knew nothing about sewage treatment," she acknowledged later.[83]

Green quickly immersed herself in issues surrounding water. By the 1980s activists had grown accustomed to gatherings in colleagues' homes; thus Green had a ready-made and enthusiastic audience. So began an organization called Heal the Bay. The problem of contaminated water "was immediately understood by everybody," she recalled. "Finding people to volunteer was so easy. I mean, people just showed up." And they sent money. "All of a sudden, a $5,000 check just arrived in the mail." Then a $50,000 check. The money went to set up an office in an industrial part of Santa Monica, to hire staff and an attorney to fight the requested waiver and to facilitate the clean-up of sewage.[84]

Los Angeles–area media outlets began to publicize the campaign. Heal the Bay hired an artist to create a logo, the skeleton of a fish. The group put together a speakers' bureau and organized a fundraiser where the Beach Boys—reared in nearby Hawthorne—performed. The group set up tables on the beach in Santa Monica to recruit members, projected images of sea life on the side of a nearby building, and held a children's march.[85] Green also contacted sympathetic politicians, including the state assembly member and longtime activist Tom Hayden, who lived in Santa Monica with his wife Jane Fonda. In 1987 Los Angeles agreed to stop the dumping and to upgrade the Hyperion facility. "You could not say no to Dorothy," recalled a member of the Los Angeles Public Works Commission.[86]

In the late 1980s Heal the Bay recruited an expert to teach Hyperion employees to create new techniques to process waste. "He went top to bottom through that organization [and] turned it into a first class operation," Green said. By that point the organization wielded immense influence in political circles, and in California as a whole. Both Democrats and Republicans sought the group's endorsements. In interviews Green took note of women's contributions in the group's success: "There's a lot of women always involved in every environmental activity that I've been involved in." She added, "We've come of age."[87]

2

SAVING THE SAN FRANCISCO BAY

In the late 1940s Sylvia McLaughlin moved to Berkeley from her hometown of Denver with her husband, Donald, a mining engineer and president of the Homestake Mining Company. Robert Sproul, chancellor of the University of California, Berkeley, had recruited Don McLaughlin to help establish the school's mining and engineering programs. Sylvia knew no one in Berkeley, so she started joining organizations: a mining auxiliary, a sewing club that created quilts for the Oak Knoll Naval Hospital, the Junior League, a Vassar alumnae group, the Faculty Wives Club, and Town and Gown, which sought to foster connections between the university and city. "I joined everything," she said later.[1]

At Town and Gown she met Catherine "Kay" Kerr, the wife of Clark Kerr, president of the University of California system, headquartered in Berkeley. Both women had children, so, like many housewives of the time, they talked

FIG. 2. Sylvia McLaughlin, Catherine "Kay" Kerr, and Esther Gulick (*left to right*) were East Bay housewives when they embarked on an improbable campaign to save the polluted San Francisco Bay from further destruction. Photo courtesy of Save the Bay.

about their families. Though the two women saw each other often at social events, they considered themselves acquaintances rather than friends. That began to change in late 1960, when McLaughlin and Kerr began a conversation that was not about their husbands or children. It was about the San Francisco Bay, which, to put it bluntly, was a mess. Something had to be done, they agreed. Thus began a partnership that lasted until the women's deaths a half-century later. With another Berkeley housewife, Esther Gulick, they soon joined forces in an extraordinary, and extraordinarily difficult—some said impossible—enterprise: to save the bay from developers, politicians, and others who over the decades had caused monumental damage to "the jewel not only of California, but of the whole nation."[2]

Little in McLaughlin's early years in Berkeley hinted that she would be anything but a traditional wife and mother, albeit one with an active and privileged life in one of the nation's premiere university towns. Her daughter and son went to private schools, and she knew many prominent business people and academics. In fact when Governor Earl Warren appointed Don McLaughlin to the University of California Board of Regents in 1951, their home on three-quarters of an acre in the Berkeley Hills became the site of many gatherings. And McLaughlin herself came from a privileged background. She was born in December 1916. Her father, George Cranmer, was a graduate of Princeton University, where future president Woodrow Wilson taught history. After college he became a stockbroker.

In 1928 he took his family, which included two sons, on a lengthy European vacation. When they returned home, the Depression loomed. Cranmer changed careers, becoming director of parks and recreation in Denver. He also designed parks and made friends with such luminaries as the famed architect Frank Lloyd Wright. Sylvia's mother, Jean Cranmer, had attended a private school in Denver, then moved to Germany to study the violin, ultimately returning home and becoming head of the Denver Symphony Society. Sylvia spent her early years in Denver, then, at fourteen, moved east to Connecticut to attend a private school. She spent time in France and majored in French at Vassar. She chose the college, she later said, because she "liked the surrounding countryside."[3]

But the Cranmer family differed in some respects from what might be viewed as the norm for that time. One of Sylvia's great-grandmothers seems to have been a Cherokee who married an Anglo man in Texas; no one talked about it, she said. Her parents supported racial integration, and her mother recruited the African American contralto Marian Anderson to perform a concert in Denver. Since the city was segregated, Anderson stayed with the Cranmers. During World War II the Cranmers housed a Japanese American man as they helped him extricate his family from an internment camp. At home, Sylvia told one interviewer, she was always treated as "one of the boys," competing with her brothers on ski slopes, in rock climbing, and on other adventures. She was thirty-one at the time of her marriage, a decade older than the median age for women of her generation. Her first child was born when she was thirty-six, and her second when she was thirty-eight.[4]

She met her future husband in 1947, when he visited Colorado to look at a mine: "[Mother] announced that Dr. McLaughlin was coming to dinner, would I go down to the Brown Palace [Hotel] to meet him?" He was teaching at Harvard and consulting on the side when he got the offer to teach at Berkeley. The couple married in 1948. He was in his fifties, divorced, and with a son only a few years younger than his new wife. "Don said we could live anywhere in Berkeley. We did not have a bay in Colorado and I just thought it was so beautiful." The home she chose featured panoramic views of the Bay Bridge that linked San Francisco and the East Bay, and of the Golden Gate Bridge. However, she soon learned that what looked pristine and beautiful from a distance, looked—and smelled—much different the closer one got.[5]

It was also much smaller than it had once been. When California gained statehood in 1850, the San Francisco Bay had spanned nearly seven hundred square miles. To the west, it opened to the ocean; to the east, the Sacramento River Delta, south nearly to San Jose, and north to Napa County. The bay had teemed with wildlife, including elk and grizzly bears, ducks, and millions of fish; it also featured forests thick with oaks, cottonwoods, and grasses as tall as full-grown men. By the 1950s, however, it had lost nearly a third of its area, much of its wildlife, and 90 percent of its wetlands.

Its length (fifty miles north to south) and breadth (twelve miles wide at some points) made it possible to overlook the diminution in size. People "looked out their windows and saw our bay, so how could it be disappear-ing?" noted the *San Francisco Chronicle* columnist Herb Caen.[6] Impossible to ignore for the four million residents of thirty-two cities in nine counties who lived near the bay was the limited amount of space set aside for pub-lic use—only 10 miles of the 276-mile coastline. There was also the nearly continual dredging and filling of the bay to make way for development and industry, and the rumble of huge trucks, kicking up dust as they rolled through city streets every day.[7]

Worst of all were the sights and smells of mountains of garbage floating in the water. More than forty entities, from cities and counties to businesses, airports, even San Quentin Prison in San Rafael, dumped their refuse into the bay. They dumped bedsprings and mattresses, unused lumber, chem-icals, dead animals, twisted steel from wrecked cars, broken crates, even an abandoned houseboat, and, of course, raw sewage—650 to 700 million gallons of waste daily—from area residents. "At night you could see the bay on fire where garbage was dumped in the shallows and set ablaze," wrote one newspaper reporter.[8] Many parts of the bay were "unsuitable for swimming because of excessively high [concentration] of human waste."[9] A *Los Angeles Times* reporter, Ronald Ostrow, grew up in the Bay Area and recalled that, as a thirteen-year-old boy during World War II he attempted to paddle a converted orange crate to a buoy in the bay. "What chilled the boy's heart was not the fear of drowning," he wrote. "It was the sight of raw sewage clinging to his paddle and fouling the water all around him." He never went into the bay again.[10]

In 1959 the Army Corps of Engineers issued a report predicting that by 2020 the San Francisco Bay could be little more than a narrow river or a shipping channel and that there would be "almost contiguous development" throughout the entire Bay Area.[11] The Corps identified 325 acres of the exist-ing bay as suitable for new filling and building. It "was a time [when] river fronts, bay fronts, ocean fronts were regarded as dumping grounds all over the country," said McLaughlin.[12]

Kay Kerr had read about the Corps of Engineers report in the *Oakland Tribune* and was concerned. She already had occasion to ponder bay issues. Her husband Clark's tenure as University of California president coincided with its emergence as one of the most celebrated higher education systems in the world. Its reputation had suffered in the early 1950s, when professors were fired for refusing to sign loyalty oaths, but it had recovered by the 1960s, largely due to the recruitment of noted scholars in many fields, the establishment of several new campuses, and the enactment in 1960 of California's Master Plan for Higher Education, which designated the University of California as a research—rather than a teaching—institution. Clark Kerr's reputation as author of the Master Plan and the University of California's elite status drew high-profile visitors from around the world. Kay often was called upon to pick up dignitaries at the San Francisco Airport and drive them over the Bay Bridge to the Berkeley campus. "I have to apologize for the appearance of the bay," she would tell her passengers. "There isn't another in the world like this but I can't see why this great big front yard is a mess."[13]

The immediate catalyst for her conversation with McLaughlin was a report that Berkeley officials planned to double the city's size by filling in an additional two thousand acres of the bay. "We've got to do something," Kerr told McLaughlin at a Town and Gown gathering. McLaughlin agreed. "I'd rather work on this than anything else," she said. Kerr suggested that McLaughlin call Esther Gulick, whose husband taught economics at Berkeley. She too had expressed concerns about the bay. "So I called Esther, and then the three of us got together out at Kay's," McLaughlin recalled.[14] As Gulick explained years later, "Kay asked me if I thought I would have time to help do something. Little did I know what that was going to mean."[15]

McLaughlin was in her midforties; both Gulick and Kerr were almost fifty. Gulick baked a batch of almond cookies to take to Kerr's because they were Clark Kerr's favorite. It was December 1960. The three women discussed the Berkeley proposal, which would create hundreds of additional homes, a golf course, and a convention center, and would adversely affect the city's Aquatic Park, one of the few shoreline areas available to the public. The trio opposed the plan but had no idea how to proceed. None of the women had any real experience in politics, though they had access to prominent academic, business, and political figures. McLaughlin had attended meetings on zoning in her neighborhood and gatherings where attendees discussed plans to build a shopping center on top of a ridge. Although she just listened in the background, those meetings introduced her to members of the city council, who were mostly "rather conservative businessmen" with regular jobs. They also were accessible to constituents. "You could just walk in and talk" to any of them. "There were no aides." It may not have been much, but it was a beginning, since council members would be among the women's first important official contacts.[16]

Existing conservation groups did have political expertise, so the three women invited prominent conservationists to a meeting at Gulick's house in the Berkeley Hills the following month. It was "a cold, clear winter night," according to one attendee.[17] Sixteen people showed up, including leaders of the Sierra Club, Save the Redwoods, and the Audubon Society. Executive Director David Brower represented the Sierra Club. San Francisco's Dorothy Erskine represented Citizens for Regional Recreation and Parks. At the end of the presentation, the conservationists said, "Yes, it's very important. We should save the bay." But they were "all too busy. They filed out and wished us luck," McLaughlin recalled. Another person put it somewhat differently: "Everybody sort of patted them on the head and said, 'well, good luck.'"[18] But the women were not deterred. As McLaughlin said, "We sat down and decided then and there that we were going to start an organization." They called it Save the San Francisco Bay Association.[19]

The conservationists may have declined direct involvement, but they offered to share their mailing lists. Erskine, who soon became a dedicated member of Save the Bay, had nearly one thousand names on her list. The other groups provided thousands more. The women decided to limit the initial mailing to Berkeley residents, since development there seemed to be the most immediate threat. They added names from their own Christmas card lists. They composed a letter describing their mission and sent out seven hundred handwritten solicitations asking each recipient to send in one dollar to join Save the Bay. They received six hundred responses; many people sent more than a dollar. Clearly they had struck a chord.[20] They named a president and convened a board of directors. Thus in January 1961 began "one of the nation's most dramatic environmental struggles."[21]

Harold Gilliam, a Bay Area newspaper reporter specializing in environmental issues, had attended the initial meeting at Gulick's home. "I was certain the three Berkeley women were too naive and inexperienced to realize the bay was as doomed as the orchards of the Santa Clara Valley.... The notion that the big-buck developers, the shoreline cities and some of the country's biggest corporations could be turned back by a few starry-eyed bay savers seemed preposterous. But I admired their idealism and kept my cynicism to myself." Gilliam, like many others, soon discovered that the women were not all that starry-eyed or idealistic and "were unperturbed by political logic."[22]

Though it would not become apparent for a while, the timing was right for a significant Bay Area conservation effort. Activists in the region had been involved in numerous campaigns over the years, including open space, housing, urban planning, freeway protests, and the effort to stop a power plant at Bodega Bay. "The freeway stuff is what got everybody excited," claimed Joe Bodovitz, an environmental activist who became closely connected to Save the Bay. "And the bay stuff came along on top of that."[23] The women might not have known much about politics, but they all had been involved in other organizations. "So we knew how it was done," McLaughlin said. From the beginning they divided up responsibilities. McLaughlin did much of the community outreach; Kerr, who had taken journalism at Stanford University, wrote press releases and coordinated strategy; Gulick, who was personally shy and retiring, took care of the administrative details.

Their first task was to familiarize themselves with the problem. "We had to give ourselves a crash course in anything that had to do with the bay," said McLaughlin.[24] Time was short, since plans for filling in the Bay and building in Berkeley were nearly completed. The women understood that their gender and inexperience could make it easy for political officials and others to dismiss them, so they sought out male experts to enhance their knowledge and credibility. Many of the experts came from the Berkeley faculty: urban planners, biologists, economists, and architects. From them the women learned about shallows and tidelands, wetlands and marine organisms. They discovered that excessive filling and dumping heated the water, which ultimately would result in hotter air temperatures, and that the garbage caused air pollution.[25]

Gulick and McLaughlin took their newfound knowledge to meetings of the Berkeley City Council and Planning Commission, where they used time allotted to the public to talk about the bay. Kerr did not attend the meetings because she lived in the neighboring town of El Cerrito, not Berkeley, and because her husband was the public face of the university. But she worked behind the scenes. Often the experts accompanied the women to meetings. Leonard Crum, a professor of economics at Harvard and Berkeley, wrote a thirteen-page paper detailing how much money railroads and other businesses would earn from Berkeley's bay plans and presented it to the council.

"We felt they were more likely to believe [the men]," McLaughlin explained. Her comment reflects the realities of the time, in which all three women, at least early on, were known by their husband's names, something that might have been a boon for Save the San Francisco Bay. No matter what they really thought, officials and others could hardly dismiss—at least publicly—the wives of such powerful men. Over time, many skeptics came to respect them. One male admirer later pondered—probably facetiously—whether the vaunted Master Plan for Higher Education actually had been the brainchild of Kay Kerr rather than her husband.[26] Another called Kerr "unrelenting and politely indomitable."[27]

None of the women looked indomitable, showing up for meetings dressed like the suburban housewives they were. "We tried to be appropriately dressed," said McLaughlin, who "usually wore a blue linen suit and a flowered hat, and possibly gloves." She wore medium-heel pumps. "I didn't want to be characterized as a little old woman in tennis shoes," she asserted, an uncharitable depiction of activist women in the early 1960s. And they chose men to head up their new organization, at least on paper. As McLaughlin said, "it was just taken for granted" that a man had to be president. Jan Konecny, a chemistry professor at Berkeley, became the first president. However, virtually everyone knew the three women—who soon came to be dubbed "the Ladies"—actually ran the organization. The president signed the letters, "but we did all the work," said McLaughlin.[28]

Despite appearances, all of the Ladies were, in fact, both indomitable and unrelenting. Gulick and McLaughlin attended virtually every meeting of the Berkeley City Council and of the Planning Department, which was tasked with making recommendations to council members about various proposals regarding the bay and other issues. McLaughlin spoke before the council. She always wrote out her talking points beforehand; the first few times "the papers shook" in her hands, she recalled. Initially the officials were merely polite, said McLaughlin: "But then, as we came to represent more and more people, they sat up and took notice." She realized early on that speakers representing only themselves tended to get short shrift, but that the politicians could not ignore a large contingent, so she always brought others along. The women also frequently dropped in to the offices of individual councilmen and planners. "We became quite friendly with some staff persons," including Jim Barnes, the head of planning, said McLaughlin.[29]

All three leaned on journalists, including the *Chronicle*'s Gilliam, to publicize their activities. They continued to expand their membership list, gaining prominent adherents, such as the photographer Ansel Adams; Adm. Chester Nimitz, commander in chief of the U.S. Pacific Fleet in World War II and a UC regent; and Newton Drury, head of Save the Redwoods League and former head of the California Division of Beaches and Parks. Caroline Livermore from Marin County joined, as did Claire Dedrick, an environmental activist on the Peninsula south of San Francisco. Dorothy Erskine's attorney husband, Morse, provided free legal advice.

McLaughlin acknowledged that the fact that all three women had connections, money, and domestic help enabled them to spend so much time on bay activities, though she denied the not infrequent allegation that they were elitists. She cited support from Bay Area labor activists and African Americans such as Ron Dellums, a local politician and later California congressman, and Byron Rumford, a state legislator.[30] She also admitted that their husbands' support facilitated their work, though Don McLaughlin occasionally complained when Sylvia came home at 1 or 2 a.m. from meetings. Every Monday for years the three principals met at Kerr's home. Sitting at her dining-room table, they plotted strategy, sent out recruitment letters, and wrote up information for inclusion in the group's newsletter.

A June 1961 newsletter warned readers, "If we don't want factories in the bay, it takes only five votes on the city council to keep them out. The citizens of Berkeley will not be given an opportunity at a public election to express their views." On July 1, 1961, Save the Bay reminded readers that "an important decision by the Berkeley City Council may be made on July 18, so early letters are needed." In October 1961 the organization sent a letter to members of the

Berkeley Planning Commission: "We wish to state on behalf of our organization that we are unalterably opposed to the extension of the city dump."[31]

By 1963 Save the Bay had 2,500 members. Each member's name and address were handwritten on a three-by-five card, different colored cards denoting new and old members and their geographical locations. The membership files were kept in a shoebox for the first few years. Many other records were kept at the university or in the women's homes. Eventually Save the Bay obtained a post office box in Berkeley, and then an office downtown.[32]

At some point during the first year, the women learned of a $5,000 grant available from the state but not yet allocated. They successfully lobbied Lieutenant Governor Glenn Anderson to give the money to Mel Scott of the UC Berkeley Institute of Governmental Studies, a well-known planning expert and president of Citizens for Regional Recreation and Parks, to conduct a study examining the history of the bay.[33]

To outsiders it might have seemed as though Save the Bay was making few inroads in its first two years, while Berkeley officials continued touting development. In 1962 they successfully pushed for an amendment to the city's tideland grant, essentially enabling them to use bay fill for any purpose they desired. McLaughlin and her colleagues were disheartened, but they redoubled their outreach efforts. "I went to schools. I went to men's organizations ... any place that we could persuade them to have us," McLaughlin said. The Berkeley Breakfast Club sometimes presented slide shows, and she took advantage of the opportunity to let attendees see the condition of the bay. Working to save the bay was "a wonderful way to meet interesting people," she added.[34]

They also ramped up the pressure on public officials, recruiting members of other community groups as allies, the League of Women Voters and garden clubs among them. "Cartons" of letters were sent to lawmakers, and people made phone calls. "We had groups in Marin County [and] down on the Peninsula," said McLaughlin. The women even used social events to promote the cause. Kerr and McLaughlin often attended regents' meetings with their husbands. Kerr played bridge with regents' wives; McLaughlin, who did not play cards, brought her household bills and wrote out checks. At one dinner, McLaughlin recalled, Lieutenant Governor Anderson, who sat on the Board of Regents, turned to Kerr and said, "Now Kay, we're not going to talk about the bay tonight." McLaughlin confessed, "We used to pester [Anderson] about ownership of the bay.... They knew we were vigilant."[35]

But the tide had begun to turn. As more people became aware of the group's activism, membership soared to more than three thousand. Pressure from activists in 1963 led to the defeat of a bill to allow corporate dredging in a part of the bay near the Marin Headlands. Berkeley officials abandoned efforts to fill the bay, decided to establish a Waterfront Advisory Committee, and appointed McLaughlin to sit on it. "[Council members] went through an agonizing reappraisal and came to see our point of view," McLaughlin said later.[36] But it was the publication that year of Mel Scott's 152-page book, *The Future of San Francisco Bay*, that provided the catalyst for wider action and helped turn "the movement . . . into a mass political uprising," as the environmental writer Richard A. Walker phrased it.[37]

The book gave Save the Bay principals everything they could have hoped for. It offered a scathing indictment of how, over a century, state and local officials and business interests had turned a priceless natural resource into "a large amount of real estate" and a "sewage disposal system." It told the story of apathy, avarice, chicanery, and greed. The degradation of the bay began in the first decade of California statehood, when hydraulic mining for gold blasted rocks and silt from the mountains north and east of Sacramento and sent the debris into rivers, reducing the size of the part of the bay that stretched toward the Sacramento River Delta.

Over succeeding decades, politicians "improvidently—and frequently fraudulently" began selling off pieces of the bay to private owners. It began with cattle barons and speculators; then came large realty syndicates, banks, title insurance companies, railroads, and manufacturing companies. In the East Bay the Santa Fe Railroad held title to "privately owned land from Richmond to Oakland" amounting to nearly 1,100 acres. Standard Oil Company owned more than 1,000 acres. In San Mateo County owners included Southern Pacific Railroad, Consolidated Western Steel, and Crocker Land Company. Lumber companies owned bay lands in Marin County. The shallowness of the bay made it particularly attractive to developers because it was easy and relatively inexpensive to fill. Eighty percent of it was less than thirty feet deep, and 70 percent less than eighteen feet at low tide. When delegates met in 1878 to write a new California constitution, the selling of San Francisco Bay was already a topic of debate. Fifty thousand acres—approximately seventy-eight square miles—had already been lost to private interests. Scott quoted one delegate, N. G. Wyatt of Monterey, saying, "If there is any one abuse greater than another, it is the abuse . . . in granting out and disposition of lands belonging to the state." Despite the concern and anger, the state continued to treat the bay as a cash register and a source of tax revenue.[38]

By the early twentieth century cities and counties had become the beneficiaries of state largesse. In the 1910s San Franciscans were raising money for an exposition to herald the completion of the Panama Canal, which meant more trade opportunities for the Bay Area. But the region needed ports to handle the additional increased traffic, so the state granted cities jurisdiction to build harbors, docks, wharves, and piers. In the 1930s the federal government funded the creation of an artificial island in the bay between San Francisco and the East Bay. It was designed for the 1939 Exposition to commemorate the completion of the Bay Bridge and the Golden Gate Bridge. Dubbed "Treasure Island," it was used by the navy in World War II.

After the war more landfill was needed for airports, both small and large. By this point the growing mountains of garbage had made the bay even shallower and thus even more conducive to development. In 1955 the state granted the city of Oakland seven thousand acres of the bay for an airport and, two years later, another three thousand acres. The definition of "airport" was left somewhat loose, to include convention facilities, restaurants, and hotels. The public might have believed that the state of California owned the bay, wrote Scott; in fact, it owned less than half. The rest was held by municipalities and private interests and "had been sold and resold many times." By 1960 the bay was "a patchwork of development rights, even further complicated by the pattern of political jurisdictions."[39]

The 1959 Army Corps of Engineers report had taken a sanguine view of additional development, but four years later Scott warned that if development remained unchecked, the bay would be "ruined for pleasure, for wildlife and as a rich source of economic wealth." He suggested that the state place a moratorium on filling and development until the legislature, which had "never applied any yardstick," could evaluate the worth of proposals for use of the tidelands and submerged lands.[40] Theoretically, two agencies oversaw bay activities. The state Lands Commission, created in 1938, managed millions of acres of tide and submerged lands throughout California, including Lake Tahoe and the Sacramento and San Joaquin rivers. The women of Save the Bay "were very critical" of it, believing it had given away too much to special interests over the years. Another group, the Association of Bay Area Governments, had been created in 1961 to coordinate land use efforts throughout the region, but it was only advisory, and members, who included city and county officials, sometimes had differing agendas. Save the Bay had "never gotten anywhere" trying to deal with the Association. Scott recommended that lawmakers authorize a single agency focused solely on the San Francisco Bay. He suggested that it be called the Bay Conservation and Development Commission.[41]

Scott's report drew significant media attention, which led to more interest in the bay and in Save the Bay. Berkeley may have abandoned long-standing development plans, but other cities and towns around the bay had not followed suit. In San Mateo County, south of San Francisco, plans were afoot to expand the San Francisco International Airport, which had been built on landfill. Scott gave Save the Bay a framework for seeking a statutory solution. The Ladies immediately shifted their focus one hundred miles east, to Sacramento and the state legislature, which held the authority to create a statewide agency to control bay development. Members of the assembly and state senate would now experience the same unrelenting pressure that Berkeley officials had received.

Even before Scott's report, the Ladies had approached state lawmakers, seeking someone to sponsor legislation to limit bay development. First up was a Democrat, Nicholas Petris, an assemblyman from Richmond. They took Petris a story by Gilliam and told him, "You're not going to have a bay anymore." Petris was sympathetic. He knew all three women and, in fact, had been an economics student of Gulick's husband at UC Berkeley. In both 1962 and 1963 he introduced bills calling for a moratorium on bay fill, but they failed to get out of committee. Lobbyists representing builders, labor unions, and cities opposed any such legislation. At the beginning, "there weren't very many that were in favor of this, other than the three ladies," Petris said later. But "they were powerful ladies."[42]

Even if the climate had been more favorable, Petris was not a prominent enough lawmaker to carry such important bills, argues the environmentalist author Rice Odell. In 1962 Petris had only been in the assembly for three years, and he was not well known outside of East Bay circles. At that point the California legislature was led "by an 'old guard'—a bipartisan, basically rural, conservative club, dominated by lobbyists for special interests," according to Odell.[43] None of them wanted controls on the bay, which might ultimately lead to controls on other bodies of water around the state. Asking Petris to carry the legislation was one of Save the Bay's few missteps, particularly since the women were not well known outside of the Bay Area themselves and thus had no clout on the state level. And they were women to boot.

No women served in the state legislature in the early 1960s, and many of the male lawmakers viewed Gulick, Kerr, and McLaughlin as little more than well-connected, bored housewives. "A lot of the business types were not crazy about the women," said Bodovitz. Kerr, in particular, came in for criticism, even among those who were on her side. "[She] could be a driving dynamo whom it was hard to resist," he said. "I mean, she'd wear you down and just keep after you, and you were either going to do it her way, or you were going to keep hearing from her."[44] As McLaughlin recalled, "We were called all sorts of names by various and sundry. That was part of the job." None of the women was intimidated; the lawmakers would soon learn, as had others before them, that the women were not to be so easily dismissed.[45]

Early one morning in 1964 Kerr drove across the Bay Bridge to San Francisco's Fisherman's Wharf, where she met Democratic state senator Eugene McAteer at his restaurant, Tarantino's. McAteer was a Berkeley graduate. Before winning a seat in the senate in 1959, he had been a San Francisco County supervisor, and he had a wide range of acquaintances and friends on both sides of the political aisle. "He was one of those tough guys who get along with power, and who therefore, on the basis of personal friendship, are able to get good bills through," writes Odell.[46]

With Kerr standing by, McAteer phoned Lieutenant Governor Anderson, who headed the state Lands Commission. "What is the commission doing about the bay?" he asked. "Nothing," Anderson replied. McAteer "hit the roof." He offered Kerr his help and told her he thought the best strategy was to create a temporary study group to investigate bay issues and problems. Study groups were common and often led to nothing in particular, but at least they represented a first step. Knowing their propensity to take charge, McAteer admonished the Save the Bay women "not to show up to hearings unless he told them to be there."[47] (Mel Scott told a somewhat different story about the origins of the study commission. He said the idea came from Governor Edmund G. Brown, who had read all of Scott's book in one afternoon and then summoned him to Sacramento.)[48] Whatever the genesis, McAteer carried the legislation creating the study group. With little fanfare, his fellow lawmakers agreed to authorize $75,000 for a study, the results of which would be presented to the legislature at the following year's session, which began in January 1965. Governor Brown signed the bill in April 1964, giving the group just over eight months to complete their work.

McAteer headed the nine-member group, which held weekly discussions around the Bay Area, open to the public. The meetings drew hundreds of participants, including high-profile conservationists such as the writer and Stanford literature professor Wallace Stegner, Ansel Adams, and area journalists. Gulick, Kerr, and McLaughlin attended every session. They brought bag lunches and later wrote up details for their newsletter. "I think they thought we were harmless," said Kerr.[49] "The whole point of the study commission was not only study, it was really broadening the group of people who might be interested in this," explained Bodovitz.[50] The strategy worked. By late 1964 Save the Bay had more than four thousand members. Virtually every Bay Area community had individuals and groups expressing growing concern about the bay.

With its higher profile and exploding membership—sometimes the organization added a hundred new members a month—Save the Bay needed a paid staff member. Janice Kittredge knew the three women because she worked for a UC Berkeley alumni group that hosted foreign visitors on campus. She earned $2 an hour to take notes at meetings, pick up mail at the Berkeley Post Office, and organize the membership roster. No longer hand-copying information onto cards, Save the Bay utilized a mailing house with an Addressograph, a machine that printed out labels.[51]

In January 1965 the study group brought its recommendations back to state lawmakers. It had spent only $45,000 of the allotted $75,000. Their sixty-four-page report included testimony, maps, photographs, and a list of thirty-four major development projects either planned or in progress, totaling 16,261 acres. By this point, interest in the bay had grown far beyond the nine-county region affected, and even beyond California. The *New York Times* and other national publications reported on developments. "Further piecemeal filling of the bay may place serious restrictions on navigation ... may destroy the irreplaceable feeding and bedding ground of fish and wildlife in the bay, may adversely affect the quality of bay waters and even the quality of air in the bay area," one *New York Times* story noted.[52] The study group recommended that the legislature create the Bay Conservation Development Commission (BCDC) and grant it the power to issue or deny permits for filling and development and to create a "master plan" for the bay: "The public interest requires creation of a government mechanism to balance competing interests."[53]

The initial BCDC would be temporary, with a four-year life span. At the end of that time the legislature would have to decide whether to make it permanent. The Ladies of Save the Bay needed to use all of their hard-won knowledge and lean on their growing number of allies to push the BCDC bill over the finish line. By this point the Sierra Club had gotten involved— the first time in its history that "the environment of a major urban area received the club's personal attention."[54] The bay was priceless and had to be preserved, declared Sierra Club leaders. By the latter half of the 1960s, Save the Bay and the Sierra Club shared officers. In 1968 Will Siri, a former president of the Sierra Club, became president of Save the Bay.

Petris in the assembly and McAteer in the senate cosponsored implementing legislation to create the temporary BCDC. Despite the growing number of supporters, success was not assured. Bay Area legislators faced continued pressure from business interests. To counter opponents, McLaughlin promoted the heavy use of media: "Perhaps the wise use of our natural resources can be 'sold' to the public with advertising techniques, on the radio, TV and in the press; with folk songs, slogans, photographs, records, and placards on buses and trains." Save the Bay wrote up flyers featuring the question "Bay or River?" and distributed them to educators, other groups, and members of the media. They mailed Bay Area voters envelopes containing bags of sand and an attached message: "You'll wonder where the water went if you fill the bay with sediment."[55]

As Petris later put it, "The public had to be galvanized. You know, fighting for control of the environment was not popular at all in those days. That was the original, basic environmental protection statute." Citizen involvement was crucial, he added, because so many entities had lined up against the proposal. Developers didn't want a unilateral agency. They thought "they could work more easily through cities." The Association of Bay Area Governments also opposed it, said Petris; they believed "[it was] much better to have it on the basis of voluntary action."[56]

Don Sherwood, a radio disc jockey, played perhaps the most significant role in selling the temporary BCDC. Sherwood was the Bay Area's most popular DJ, with a morning drive-time program on KSFO. A thin, "almost gaunt man," he lived on a houseboat in the Marin County town of Sausalito. "He loved to pull stunts, loved making fun of pious people in positions of power," and he "loved to stir things up." Sherwood was acquainted with McAteer, who told him about the legislation. Many mornings, while on the air, Sherwood phoned McAteer to ask "What's going on with the bill?" Then he asked, "If people in the audience want to help, is there something they can do?" And McAteer would respond, "Write a letter to your [state] senator or assembly member." Lawmakers were inundated with mail. In fact they received more mail on the proposed BCDC than on any other piece of legislation that session. Sherwood even phoned Brown in the Governor's Mansion early one morning to talk about the bill.[57]

The hearings to debate the BCDC that began in spring 1965 were packed. McLaughlin, Gulick, Kerr, and hundreds of other Save the Bay partisans organized car caravans and chartered buses from the Bay Area to Sacramento for hearings. Sometimes their children accompanied them. McLaughlin wrote to lawmakers and sent them clippings from the growing number of national publications covering the debates, and from health professionals. During some legislative sessions she sat in the back of the chamber. When lawmakers walked by or sat down near her, she handed them information while remarking, "Sir, I have something for you."[58] Constant pressure and media attention undoubtedly led some lawmakers who might have opposed the legislation to vote yes.

The bill started in the assembly's Governmental Efficiency Committee and passed out by one vote. Petris was aided by his Democratic colleague John T. Knox of Berkeley in getting the bill through the lower house. In the state senate, "McAteer came to the rescue," Petris said later. After that body voted narrowly to approve the measure, Brown signed it in September 1965. The commission had an annual budget of $230,000 and a governing body of twenty-seven members, representing federal, state, and local governments. Bodovitz was named executive director. "It was an ingenious innovation," declared Gilliam, "the first agency in the country to limit regional development, with unprecedented power over cities, counties and private developers." Creation of the BCDC—even as a temporary agency—"demonstrated the power of grassroots action in a democracy and provided a model for emulation elsewhere."[59]

The new commission immediately implemented the permit system for bay development. People quickly realized, said Bodovitz, that "there was a serious agency that was going to protect the bay, so don't waste your time and money trying to get a permit to do something that isn't going to happen. . . . One of the remarkable things about the Save the Bay movement and BCDC was how quickly public attitudes changed." Previously "garbage companies routinely staked out areas of the bay to dump garbage. They knew that day was gone."[60]

At the beginning of their journey, Gulick, Kerr, and McLaughlin had needed their husbands' clout to gain any traction or attention for their cause. By the mid-1960s they nearly equaled their husbands in influence and could stand on their own. This was fortunate, because two of the three husbands, Clark Kerr and Don McLaughlin, saw their stature diminish significantly by this point. A political tsunami was beginning to wash over the country, sweeping away many long-term leaders of the establishment. In their place would be "outsiders," including Ronald Reagan, soon to be governor but until 1966 viewed in many quarters as a washed-up actor. Engaged citizens like Gulick, Kerr, McLaughlin, and others were quickly becoming the future of political activism.

The problems for Clark Kerr and Don McLaughlin began in September 1964 in Berkeley, as Save the Bay was closely monitoring the work of the study group. Dozens of politically engaged students returned that fall from "Freedom Summer" in Mississippi, where they had experienced harassment, threats, and violence as they tried to register African American voters. They returned to Berkeley to find a campus still steeped in the tradition of in loco parentis, where professors and administrators were the ultimate authority figures and students could be disciplined or expelled seemingly for any infraction. But the students had changed and no longer accepted the status quo. When administrators refused to allow them to set up card tables around the campus perimeter to collect money for civil rights campaigns, protest erupted. The next three months saw nearly constant turmoil and the beginnings of the Free Speech Movement. Thousands of students sparred with authorities. They stood on police cars, occupied administration buildings, and goaded police to arrest them. When police from Bay Area cities became overwhelmed, authorities began wholesale arrests of protesters.

Forty-five years later Sylvia McLaughlin remembered that time clearly. At one point Clark Kerr called the McLaughlins at 3 a.m., asking for advice. "Don went over and met with students. We became good friends with some of them." She also remembered the eye-stinging presence of tear gas when she went to downtown Berkeley or picked up her children from school.[61] Administrators were caught between the students' anger and rage from older people, who thought the protesters should be thrown in jail and left there. By 1965 Berkeley's chancellor, Edward Strong, had lost his job. Clark Kerr and Don McLaughlin would soon follow. In November 1966 Brown lost his bid for a third term to Reagan, who campaigned on the ingratitude of unruly college students, who, he declared, were fortunate enough to get a virtually free education, courtesy of the state of California. In one of his first actions, he fired Kerr as University of California president. McLaughlin's term as regent was up in 1966, and Reagan did not reappoint him. By 1966, as the temporary BCDC began its work, the Vietnam War was tearing America apart. Citizen activism was on the rise everywhere.

At this point Save the Bay had nearly ten thousand members and counting. As the catalyst for "the first major revolt against the dominant postwar mind-set of unrestricted development," as journalist Gilliam described it, the group drew many volunteers.[62] These included senior citizens, young people—including Campfire Girls—and conscientious objectors from the Vietnam War, some of whom became paid staff members and officers. The San Francisco Foundation gave Save the Bay a $5,000 grant to make a film about their efforts. Over the next three years, as the organization geared up to battle for a permanent commission, membership soared to fifteen thousand, then to nearly twenty thousand. Dues were still $1. By the latter part of the 1960s, political candidates were seeking endorsements from environmental groups. Pete McCloskey, a liberal Republican, later attributed his successful 1966 congressional campaign in Santa Clara County to environmental supporters.[63] "Books were written and television programs made about the threatened bay," writes Walker. It "became a wildly popular cause.... It [the rising profile of environmental activism] all goes through Save the Bay."[64]

Meanwhile the temporary BCDC enforced "a tight moratorium on bay filling." By the end of its tenure, the agency had issued fifty-six permits for only 370 acres of fill, mostly for the expansion of Oakland's airport. Commissioners recommended filling only for uses "providing substantial public benefits and treating the bay as a body of water, not as real estate." But they provided some latitude. Ports, along with "industries that require deep-water shipping, airports which cannot be located away from the bay, waterfront freeways under the same circumstances, publicly-owned recreational facilities and commercial facilities such as restaurants and hotels, if they provide new public access to the bay," would be allowed.[65]

More evidence of Save the Bay's enhanced clout came in 1968, when a corporation named Westbay Community Associates sought to build a port, restaurants, hotels, and other commercial and industrial buildings on more than ten thousand acres along twenty-seven miles of the San Mateo County coastline south of San Francisco. One proposal would have leveled a small mountain in the city of San Bruno. The Westbay principals were not bit players, but powerful and prominent men, among them David Rockefeller. When San Mateo objected to the company's plans, the state of California filed suit against Westbay, which had claimed that hundred-year-old grants of submerged lands gave it the right to develop. The state attorney general argued that such grants were illegal. Save the Bay sought to intervene on the state's side and was permitted to do so, marking "the first time an environmental organization was granted the right" to participate. It "proved that environmental groups had standing to appear in court and argue what they perceived to be in the public interest," said Gulick. "It was extremely important because . . . the lawsuit could not be settled without our agreement."[66]

In early 1969 the BCDC presented its San Francisco Bay Plan to Governor Reagan and the legislature. Environmentalists knew the fight would be fierce, despite the rising profile of environmentalism in general and the protests and outrage over disasters such as the massive Union Oil spill off the Santa Barbara coast in late January and early February 1969.[67] Many city and county officials, corporate executives, and developers remained virulently opposed to controls on San Francisco Bay, fearing that closing off one area could lead to other efforts throughout the state. High-profile opponents included the city of Oakland and Leslie Salt, which owned forty-six thousand acres of salt ponds and wanted the ability to sell off its lands for development. Lawmakers who received campaign donations from business entities seemed reluctant to cross their patrons. "There were certain legislators who we knew were sympathetic to some of the developers," McLaughlin said. "The developers had hired high-paid lobbyists."[68]

Legally the conservation groups could not engage in political lobbying without losing their tax exempt status. In fact the IRS revoked the Sierra Club's tax exemption in 1966 for publicly lobbying against a dam in the Grand Canyon. In 1968 a coalition of Bay Area environmental groups, including the Sierra Club, Save the Bay, and the Greenbelt Alliance, created a lobbying organization, the Citizens' Alliance to Save San Francisco Bay. Together the groups had more than sixty thousand members. Citizens' Alliance aimed to raise money and recruit twenty thousand people to write letters and to flood the Sacramento hearings with constituents.

Once again McLaughlin realized that publicity would be the key. She wrote and distributed materials for teachers and students, participated in conferences and public meetings, and put together a slide show on the bay. The group also sent newsletters to its many members. Dorothy Erskine, who served on the BCDC board, condensed material from the commission reports to four pages before sending it on to Kay Kerr, who whittled the information down to one page to be published in Save the Bay's newsletters. "There would be some lines about a particular issue, accompanied by a photo or cartoon," McLaughlin explained. One cartoon depicted Little Red Riding Hood as the San Francisco Bay standing at the foot of a bed containing the wolf, depicted as a huge bay dredger with snapping jaws. The caption: "The Legislature Can Save the Bay and Not the Wolf."[69]

Technically the legislation under consideration represented a continuation of the original McAteer-Petris Act. By this point few lawmakers were willing to go on record leaving the bay open to unbridled development with no controls, so environmental groups knew they could expect authorization for at least some permanent role for the BCDC. The question was, how much? Some cities sought exemptions. Environmentalists wanted the agency to have control of the shoreline, but how far out? One thousand feet? One hundred feet? Sadly, McAteer was no longer available to guide the legislation through the senate; he had died of a heart attack in 1967. Ten bills initially had been introduced for the 1969 session, but only four "received serious consideration": one in the assembly, authored by John T. Knox, and three in the Senate. Petris, who had moved to the senate and gained much more clout, authored one of the three. The others were by Richard Dolwig, a Republican from San Mateo, and Milton Marks, a Democrat from San Francisco. Environmentalists favored only Knox's and Petris's bills, viewing the other two—particularly Dolwig's—as too weak or even designed to quietly undermine the BCDC. Dolwig, they believed, was far too friendly to corporate interests.

In the assembly, Knox's bill garnered widespread support, particularly after a large group of environmentalists "laughed at, hissed, and shouted down" developers in committee hearings. It moved through two committees and onto the floor, where the full membership voted 55 to 9 for passage. But the outlook for passage of a "strong preservation bill seemed dim" in the state senate. To enhance prospects there, another organization emerged, dedicated to promoting the BCDC and to publicly shaming opponents of strong bay controls. Janet Adams and Claire Dedrick knew what they were doing. Both were environmental activists, members of Save the Bay, and public relations experts. In April 1969 they created Save Our Bay Action Committee (SOBAC). The group's "basic strategy was to . . . concentrate on several simply understood issues." It promoted the notion that the bay belonged to everyone, and it offered strong support for the Knox bill.[70]

If every story needs a hero and a villain, SOBAC chose Richard Dolwig to play the latter role. The first SOBAC ad appeared the day after Dolwig gaveled open a hearing on the proposed legislation in his Governmental Efficiency Committee—long deemed the burial ground for environmental legislation. Hundreds of Bay Area conservationists, including Gulick, Kerr, and McLaughlin, rode chartered buses to Sacramento, jammed the hearing room, and—in an unprecedented move—forced Dolwig to move the meeting to a larger room. Adams and Dedrick created "38,000 deep blue and chartreuse bumper stickers bearing the slogan 'Save Our Bay.'" They circulated petitions, garnering thousands of signatures. They placed full-page ads in area newspapers urging readers to show up in Sacramento for hearings. One ad read, "TODAY, DEMAND of Senator Richard J. Dolwig . . . the removal of his own bill . . . and any other Bill designed to exploit San Francisco Bay and its shorelines." Another stated, "Pollution has reached a level as to seriously damage plants and animals in the Bay area. . . . WIRE! WRITE! PHONE!—to Demand a Halt to 122 years of Destruction of San Francisco Bay."[71]

They sent Dolwig's constituents envelopes containing sand and the message "Fill the Bay, with Dolwig." SOBAC worked with other environmental groups to send a petition with nearly eighty thousand signatures supporting the BCDC to Dolwig's office. More "than 400 letters advocating a strong BCDC bill arrived at his office each day."[72] Shortly thereafter, Dolwig quietly announced that he'd had a change of heart; he now backed a strong bay bill. He claimed that staff conversations had led to his reversal, but Petris credited environmentalists for shining a spotlight on Dolwig, asserting that they "did a very skillful job. . . . They flooded the district. They did a fantastic job."[73]

Soon another obstacle was removed. Hugh Burns, a Fresno Democrat widely perceived as unsympathetic to the BCDC, was president pro tem of the senate when debate began. As such, he had the power to decide which committees received legislation and to hold up legislation altogether. In spring 1969 his colleagues ousted him from that position and replaced him with Howard Way, a Democrat from Exeter and a strong supporter of the BCDC. One colleague later acknowledged that "the BCDC struggle led to Burns' defeat." Others, including environmentalists, saw Way's elevation as heralding "a new era in which the Senate would be more responsive to public interest legislation."[74]

By this point Knox's measure had become the only bill under consideration. Soon after becoming pro tem, Way "used his personal influence to speed progress on the bill." He assigned it to the Local Government Committee rather than the much tougher Governmental Efficiency Committee. It then went to the Finance Committee, then the entire senate. When it appeared that lawmakers from Southern California might try to sabotage the effort, SOBAC hurriedly unleashed a campaign there, placing an ad in the Los Angeles Times. The San Francisco Bay "belonged to all Californians," it read, and "it was one of the state's unique natural resources."[75]

All three principals of Save the Bay attended virtually every hearing. McLaughlin even testified at a few. She rehearsed her talking points on the bus from Berkeley to Sacramento, she recalled: "I always tried to return to my seat quickly [after testifying], so I wouldn't be asked questions." Once, she failed to move fast enough and a lawmaker asked her about a particular fish. Fortunately she was ready, having studied that topic during her preparation.[76] In the last senate debate before the vote on the final bill, McLaughlin "was sitting there with Esther": "I think we were holding hands nervously as they counted the votes." Radio and television stations covered the session, and anxious spectators packed the senate chambers and jammed the hallways outside. At the end, the measure passed easily, 24 to 10.

Governor Reagan represented the final hurdle. Environmentalists feared that, ultimately, he would side with developers and insist on weaker protections; he had once been quoted saying, "Once you've seen one redwood, you've seen them all." He also "expressed reservations" about giving BCDC jurisdiction over the shoreline, taking away regional control. On the other hand, he had offered some encouragement and even mentioned the BCDC in a speech before the legislature, And he had appointed environmentalists to his cabinet. They included Caroline Livermore's son Norman Jr., called "Ike," as resources secretary, and the landscape architect William Penn Mott of Save the Bay, as the director of parks and recreation. In August 1969, amid some fanfare and much jubilation, Reagan signed the bill. Gulick, Kerr, and McLaughlin attended the signing ceremony in Sacramento, as did their former antagonist and now ally Richard Dolwig, and the bill's authors, Knox and Petris.[77]

After more than eight years the women of Save the Bay could finally claim success. McLaughlin credited the large community of environmentalists, but also noted that the measure to make BCDC permanent had been the only environmental bill before state lawmakers in 1969, enabling supporters to focus all of their efforts on a single piece of legislation. The BCDC legislation was not perfect. It included exemptions for a few cities, such as Emeryville and Albany, that already had development plans in the pipeline. And it gave the BCDC jurisdiction over one hundred feet of shoreline. There would continue to be conflict between developers and environmentalists, and between regional interests and the state. Water pollution would not be eradicated. But the bay would no longer be a dumping ground for cities' garbage nor the source of easy riches for developers.

Succeeding years saw much more environmental legislation. In 1970 President Richard Nixon signed the National Environmental Policy Act, and the country celebrated the first Earth Day. In November 1972 California voters approved Proposition 20 to create a Coastal Commission to monitor the state's entire coast; the BCDC had served as a model. The commission initially had a four-year life span; in 1976 voters made it permanent. In the years that followed "there was a gradual proliferation of local groups like the Committee for Green Foothills, Save Mount Diablo, San Bruno Mountain Watch, the Greenbelt Alliance, San Francisco Tomorrow and the Marin Agricultural Land Trust," writes Gilliam.[78]

Over the next decades Gulick, Kerr, and McLaughlin continued to be celebrated in environmental circles. In 1972 they were featured in Odell's book, *The Saving of San Francisco Bay: A Report on Citizen Action and Regional Planning.* They were interviewed by journalists, won awards from numerous organizations, and were in demand as speakers, traveling throughout the country. In 1974 Edmund G. Brown's son Jerry won the first of what would be an unprecedented four terms as governor. He named Claire Dedrick of

Save the Bay and Save Our Bay Action Committee as resources secretary, responsible for overseeing land use policy for the state. She was the first woman to hold the job and went on to serve on the California Public Utilities Commission.[79]

In 1977 the Westbay Community Associates lawsuit was finally settled on terms favorable to environmentalists. Afterward McLaughlin was introduced to David Rockefeller. "He held out his hand and said, 'you win.'"[80] By the 1980s all three women were in their seventies. Through the years they had continued to meet around Kerr's dining room table every Monday to devise strategy, though McLaughlin sometimes skipped the meetings as her children grew older and needed more attention.

Meanwhile the BCDC began its effort to reverse a century of bay destruction. Between 1974 and 1983 the agency reviewed and acted on applications for 232 large projects valued at more than $1.8 billion. Commissioners denied some and approved others, with the condition that developers ensure public access to the bay. By 1983 there were four hundred acres of new bay, only seventy-six acres of new fill, and twenty-nine additional miles of public access along the shoreline. However, the agency could not eradicate all of the bay pollution from oil refineries, irrigation, pesticides, and fertilizer.[81]

In 1985 and 1986 the three women sat down for several interviews with Malca Chall of UC Berkeley's Regional Oral History Office, to discuss the movement they had started. In her introduction Chall wrote, "Without their concerted skills as leaders, organizers, researchers, writers, fund raisers, and arm twisters there would be no Save the San Francisco Bay association, no Bay Conservation and Development Commission as we know it, few court cases on behalf of the public trust—in short, there would be a diminished bay, diminished public access, and diminished public awareness of the value of San Francisco Bay."[82]

"We never thought we'd still be at this twenty-five years later," McLaughlin admitted. Both Charles Gulick and Don McLaughlin had died in 1984. Though common practice was to interview one subject at a time, the women insisted on being interviewed together. "We've always been a team," Kerr explained. Prior to the interviews, the women sent Chall a list of questions they thought she should ask. In the brief biographies accompanying the interviews, each of the three women listed "housewife" as her primary occupation, although they also described themselves as "activists," "volunteers," and "environmentalists." During the several sessions they reminisced, laughed, and talked over and corrected each other, as well as Chall. They asked to have the transcripts typed up and distributed so they could correct the record. Each possessed, it seemed, a photographic memory of every event, issue, and person involved in Save the Bay's campaigns. Transcribing the interviews was made difficult by the conversations the three women had among themselves as the interviewer waited for their responses. Additionally, by 1986 their voices and intonations were so similar, they all sounded alike. Speaking for herself and her two colleagues, Kerr said, "By training and temperament, we believed that a persuasive rather than adversarial approach was most effective. We also realized that knowledge was essential before taking a position."[83]

In 1990 the *Los Angeles Times* reporter Ron Ostrow returned to San Francisco, the site of his nightmarish childhood experience with bay sewage. "Progress indeed has been made," he wrote. One hundred miles of shoreline was now accessible to the public. Raw sewage no longer clogged the bay "as a matter of course." Pollutants had decreased by 80 percent, though toxic chemicals were "still flushed into the bay," mostly due to runoff from city streets and rooftops. "Most encouraging of all, there is a pervasive concern about the environment that was virtually non-existent 30 or 40 years ago."[84]

In 2007 McLaughlin again sat down for an oral history interview, this time alone. Gulick had died in 1995 at the age of eighty-four. Clark Kerr had died in 2003, and Kay Kerr was no longer as active as she once had been. Both she and McLaughlin were then in their nineties. McLaughlin still belonged to a dizzying array of groups, including People for Open Space, Save the Redwoods League, National Audubon Society, and the East Bay Conservation Corps. She focused much of her attention on parks. At the time of the interview, she was involved in a campaign to save a grove of oak trees on the Berkeley campus. Donors wanted to replace the trees with a concrete building to house athletics. Said donors had "misplaced priorities," McLaughlin declared. "I don't think they show very much concern for the neighbors." She had demonstrated her own concern by climbing a tree and remaining there for several hours.[85]

She discussed being an activist at a time when women of her age and class were expected to stay at home. Over time she had come to a different understanding of the role of homemaker. "Women take care of the home," she told her interviewer, Ann Lage. "Well, I've come to realize that I'm a homemaker for my home, my garden, my neighborhood, my community, and the larger community of this region, the state, the planet." She added, "Who knows—maybe the environmental movement helped the woman's movement come along."[86]

In 2011 San Francisco's public television station KQED aired a four-part series, *Saving the Bay: The Story of San Francisco Bay*, narrated by Robert Redford. McLaughlin was the only Save the Bay founder still alive. Kerr had died earlier that year at the age of ninety-nine. McLaughlin and Gilliam reprised the story of the initial meeting at Gulick's home in January 1961. Gilliam recalled his initial impression of the women as "noble idealists." Looking back, he said, the conservation movement had been at a crossroads, though few could have predicted where and how far it would go in just a few years. By 2011, "environment" had become the word of choice. "It turned out that everybody wanted to fill the bay—everyone but the people." Save the Bay's genius was to help members of the public realize how much they had lost, Gilliam said, and how much more they stood to lose, unless they took action.

Sylvia McLaughlin died in January 2016 at the age of ninety-nine. Today, Save the Bay is headquartered in Oakland. It has more than fifty thousand members and many more volunteers, who spend hours each month cleaning up trash, reseeding, and replanting. The bay is approximately 150 square miles larger than it was in 1961. It is ringed with shoreline parks, including three hundred miles of bike paths and walking and hiking trails. Eastshore Park, with 8.5 miles of shoreline access from Berkeley to Richmond, has been renamed McLaughlin Eastshore State Park. Save the Bay also helped to secure passage of the state's first wetlands protection law, and more than 100,000 acres of wetlands in San Francisco Bay have been restored. Among its recent successes: helping to pass a regional, then a statewide ban on plastic bags. Save the Bay "is a great American success story," Gilliam declared. "Democracy at its best, a grass-roots action that overcame overwhelming odds." The "ripple effects went far beyond the bay. They went across the country and around the world and into history."[87]

3

Fig. 3. Kathleen Goddard Jones loved to hike on the pristine Nipomo Dunes. When she learned that PG&E planned a nuclear power plant there, she embarked on a crusade to stop the giant utility. Source: the *Tribune*. © 1985 McClatchy. All rights reserved. Used under license.

THE DUNE LADY

Kathleen Goddard Jones

Kathleen Goddard Jackson was in her midfifties and the mother of five, living in the central coast town of Paso Robles, California, in early 1963 when she saw an item in a local newspaper announcing that Pacific Gas & Electric Company had purchased 1,100 acres of the Nipomo Dunes. The utility planned to build a nuclear power plant on the site. The eighteen-mile stretch of land straddled two counties, southern San Luis Obispo and northern Santa Barbara, and spanned an area from Pismo Beach in the north to Santa Maria in the south.

The proposed plant would be built in San Luis Obispo County. Situated about halfway between Los Angeles and San Francisco, the county spans 3,600 square miles, from the Pacific Ocean to the edge of the San Joaquin Valley and from Monterey County to Santa Barbara County. Except for a dozen or so miles south of the city of San Luis Obispo, where it hugs the coastline, Highway 101 runs inland, north to south, loosely following the old El Camino Real from the Spanish era. Beach towns lie to the west, with inland hills, valleys, and mountains to the east. The county features a wide variety of microclimates: cooler coastal weather and warmer inland, particularly in the northern part of the county, with its proliferation of chaparral, oak trees, almond and olive groves, and miles of vineyards.

The city of San Luis Obispo grew up around its mission, built in 1772, the fifth of twenty-one established by Franciscan friars. The county's most notable landmark is Hearst Castle, a fabled 110-room hillside estate situated on 130 acres outside the town of San Simeon in the far northwestern part of the county. From the 1920s to his death in 1951, the publishing magnate William Randolph Hearst spent much of his time living in the castle with his mistress, Marion Davies.

In the 1960s the county was still sparsely populated, with approximately ninety thousand residents clustered mostly in a handful of cities: San Luis Obispo, Arroyo Grande, Atascadero, Morro Bay, Paso Robles. The county's economy depended mostly on agriculture—fishing, farming, and ranching—and on small businesses, many affiliated with farming and ranching. Agriculture was a major focus of California Polytechnic State College (now university), the region's only four-year college.

For many county residents, the news of a possible nuclear power plant was heartening; it meant jobs and tax revenue. For Jackson, however, the news was most distressing. To her, the sculpted dunes were a nearly sacred place. She knew the terrain intimately, having hiked it dozens of times. Often she was alone, but she also led Sierra Club hikes in the area. It was, in essence, her church, the place she went to gather her thoughts, find solace and serenity. She memorized the names of birds and fish that lived in and around Oso Flaco Lake, the entry point to the Dunes, as well as the flowers, shrubs, and trees dotting the constantly shifting sands that rose in places to five hundred feet before dropping off to reveal a breathtaking view of the ocean. There were least terns, snowy plovers, purple sage, sticky monkey, scarlet Indian paintbrush, bright orange California poppies, eucalyptus trees. The ecosystem also contained many endangered species.

In the early 1960s Jackson did not yet know the Dunes' colorful history, though she would come to know it better than virtually anything or anyone in her life. Members of the Chumash tribe had lived in the area until European settlers arrived in the eighteenth century. The Spanish explorer Gaspar de Portolá had passed through in the late 1760s on his way north to Monterey.

According to legend, he killed a "skinny bear," hence the name Oso Flaco. In the early 1920s the film director Cecil B. DeMille used the Dunes as the setting for his silent film classic *The Ten Commandments*. A few years later, a group of "bohemians" discovered the area and constructed a commune of sorts, designed to be a place of "solitude and enlightenment."[1]

The commune's nominal leader was Chester A. Arthur III, known as Gavin. Before settling on California's Central Coast, the grandson of the former president had lived in Ireland for a time, where he procured weapons for the Irish Republican Army. Arthur named his dunes community Moy Mell, Irish for "pastures of honey." The group came to be popularly known as "Dunites." Other members included George Blais, "a reformed alcoholic who preached nudism, vegetarianism, and sleeping under the stars"; Elwood Decker, an artist and devotee of Eastern philosophy; a Hindu mystic named Meher Baba; and Dunham Thorpe, a Hollywood scriptwriter and publicist for the actor Joan Crawford.[2]

Women lived there as well, including, for a time, the Irish poet and writer Ella Young. Thorpe's young daughter spent six years at the commune with her parents, before moving to Berkeley. Frequent visitors included the writers John Steinbeck and Upton Sinclair and the photographers Ansel Adams and Edward Weston. Most members of the group lived in small cabins, constructed of scavenged wood. They subsisted mostly on Pismo clams, and vegetables and fruit from their garden. For a brief period the Dunites even published their own magazine, *Dune Forum*.[3]

The Dunites had disbanded by the time PG&E bought the land. Earlier developers had sought to build there, but the shifting sands made it impossible. Oil had been discovered on a portion of the Dunes in the 1940s, and oil companies, including Unocal, owned much of it. In fact PG&E had purchased the land from Union Oil. It had been zoned M2, "the stinkiest, noisiest, dustiest type of industry, the most polluting, the most undesirable imaginable," Jackson recalled.[4]

PG&E was the largest utility in California, providing electricity and natural gas to two-thirds of the state's residents, mostly in the northern half of the state. In fact the company had erected what became a natural gas–powered plant in San Luis Obispo County in the 1950s, in the beach town of Morro

Bay. Along with other utilities, beginning in the early 1960s it had begun to tiptoe into nuclear energy. It had the backing of politicians, including President Dwight D. Eisenhower, who promoted nuclear energy for peacetime use. The Atomic Energy Act of 1954 encouraged investor-owned companies like PG&E to build facilities, but corporate officials demanded federal backing to minimize financial liability. PG&E envisioned California as "a base, or ground zero for a new age of energy production." Since the reactors would have to be cooled by water, most proposed plants would be sited along or near California's 1,100-mile coastline, more than 60 percent of which was owned by private interests and individuals. Other California utility companies, including the Los Angeles Department of Water and Power and Southern California Edison, planned to build their own nuclear facilities.[5]

Most California residents supported nuclear power, which was touted as cheaper, cleaner, and more reliable than energy from fossil fuels. Additionally, "electric power had become a central element of the liberal platform to raise America's standard of living." Nuclear-powered electricity was to be "the basis of a new American culture" that saw consumerism as a ticket to the middle class and a rung on the ladder to social success and status. It was sold to suburban housewives as a way to make their lives easier with appliances.[6] It even gained cachet in popular culture. In 1956, for instance, Walt Disney created the cartoon *Our Friend the Atom*.[7] PG&E planned eventually to build nearly a dozen facilities in California.[8]

Despite the company's political and economic clout, it faced some opposition. In the mid-1950s the Atomic Energy Commission and California Public Utilities Commission had given PG&E the go-ahead to locate its first large, commercially viable facility at Bodega Bay. A small Sonoma County town about fifty miles north of San Francisco, Bodega Bay is perhaps best known as the setting for Alfred Hitchcock's film *The Birds*.[9] Businesses, local government officials, and most residents supported the proposed plant. It was to be built on Bodega Head, a rocky, finger-like promontory that shelters the bay below.

But Rose Gaffney, whose father had purchased nearly five hundred acres of land in the area in the 1800s, refused to sell. "I was dead set against locating an atomic power plant there," she told a reporter in 1971. "I raised a helluva lot of noise about the damn foolishness of the whole deal."[10] Nonetheless PG&E used eminent domain to purchase sixty-five acres from Gaffney at $1,000 an acre. Furious, she contacted the Sierra Club, which declined to intervene; the group had no stated position on nuclear power, she was told. Additionally the club was trying to stop the building of dams in wilderness areas, and some members saw nuclear energy as an alternative to dams. As one member later commented, "The Sierra Club can talk about scenic beauty ... but not about public safety."[11]

But some Club members disagreed with the hands-off stance. They included a young lawyer named David Pesonen, who believed that too many people "had abdicated control of their lives and [had ceded] political power to a small and elite group of nuclear experts." He wrote investigative pieces for the *Sebastopol Times* and recruited UC Berkeley faculty to lecture on dangers posed by nuclear power. Other groups and individuals soon got involved, including the Bay Area activist Jean Kortum. She picketed PG&E's San Francisco headquarters and organized letter-writing campaigns. When seismologists discovered that the San Andreas Fault ran through the proposed site, PG&E was forced to abandon its plans.[12]

Short of finding another major earthquake fault, however, PG&E's immense clout and seemingly bottomless supply of funds made stopping the proposed plant on the Nipomo Dunes appear the longest of long shots. But Kathleen Jackson was not deterred by others' power and money, and she did not fear those in positions of authority. To most people who met the diminutive, freckled redhead, she must have seemed to be just another housewife who loved nature and liked to hike. In fact after the first meeting, one PG&E official quipped, "What can she do? One woman?"[13]

They would quickly find out. So would leaders of the vaunted Sierra Club, long led by powerful men who set the group's goals and agenda. For the nation's oldest and premier conservation group, the fight over the power plant would fuel an existential crisis. It would set off larger debates within the Club about development and growth and challenge the power of long-standing conservation leaders to set the nation's environmental agenda. All because a middle-aged woman with a winning smile and a spine of steel decided to devote her life to saving the Dunes.

Jackson may have looked to all the world like a traditional wife and mother, but the woman known to friends and family as "Kathy" had her own colorful past. In fact she might have been right at home with the Dunites. Born Kathleen Goddard in July 1907 in Sacramento and reared in Santa Barbara, she took early to hiking and camping, "going to the higher peaks" even before her teens. At Santa Barbara High School she became the first freshman to win the Silver Barry Oratory Cup. "Talking is one of the things that has come easily to me," she quipped to an interviewer years later. After high school she spent a year at Santa Barbara Teachers College, working in the library to earn money. In June 1926 two older women friends invited her to join them on a trip to Europe; she immediately accepted. She was nineteen.[14]

"I felt like a bold, bad pirate," she wrote to her mother on the first leg of her journey. It took her via train from California to New York, and then on a steamship to Europe. Crossing Nebraska, she wrote of "meadowlarks and endless, undulating green." Sunset brought "knobby hills that made grotesque silhouettes." The ocean crossing took seven days. In England she stayed out late and described the pale moon and the light from its "egg-shaped face." In Italy she wrote back to her mother, who had admonished her about drinking wine, "Wine is as water over here and is served with the meal without extra charges." She added, "All of the women here smoke." In Paris she wrote about prostitutes who appeared unkempt during the day but "blossomed" at night, and in Capri she described barhopping with a friend until 4 a.m.: "I feel as though I were tolling a knell." At some point she became separated from her friends and traveled mostly alone. "If you knew the places I'd been . . . at all hours," she wrote her mother, "if you knew half my adventures, you'd be pawning your jewelry and sending out a search party." After nearly three months abroad, she sailed back to New York.[15]

It was late September 1926 when Goddard enrolled in Mills College, a women's college in Oakland. There she met Cedric Wright, a music professor, writer, and Sierra Club activist, who invited her to parties in San Francisco. He introduced her to Ansel Adams and Aaron Copeland. Smitten with Goddard, Wright proposed marriage. She demurred, citing the ten-year age gap. Despite the rebuff, Wright and Goddard would remain lifelong friends.

A short time later, at a San Francisco rug store, she met and soon married Ali Shirazi Parvaz, an Iranian pilot who had flown for the Allies during World War I. The couple spent several years living in Washington DC, India, Burma, and Iran, before moving to New York. The marriage had begun to fray by then, and in 1931 the couple separated and Goddard took a job at NBC radio network. She started in the steno pool but worked her way up to the publicity department and occasionally appeared as a guest on talk shows.

In 1938 she appeared on the radio show *Let's Talk It Over*, hosted by Alma Kitchell, who invited her to discuss living in the Middle East. Women were treated well in Iran, Goddard said. They wore European clothing, worked in business, and "crusaded for social change." They also had equal rights when it came to marriage and divorce. As an aside, she opined that children of all cultures should play together as "the best way to teach democracy to new generations." On another program, discussing architectural design in housing, Goddard said her ideal home would contain a bathroom for each person, with a light over each bathtub for reading. Such designs were unfeasible, she acknowledged, because virtually all architects were men.[16]

Goddard divorced Parvaz in 1940 and returned to Santa Barbara, where she joined the Sierra Club under Wright's sponsorship. She recalled, "At last I could pick up the threads of something that had been sleeping in my heart all those years."[17] In 1945 she married for a second time. Duncan Jackson was a well-to-do businessman and amateur musician, active in the Santa Barbara Symphony. Kathleen was then in her late thirties. Unable to have children themselves, the Jacksons decided to adopt—five children within just a few years. But domesticity by itself proved unfulfilling, so Kathleen began to involve herself more actively in the Sierra Club. In 1949 she joined a two-week backpacking trip through the Sequoias led by Wright and the Club's executive director David Brower. Santa Barbara had no official chapter, so in 1952 she helped to create the Los Padres Chapter. She soon became its chairperson as well as editor of its newsletter, *Condor Call*.

Jackson's leadership status at the local level, plus her media experience, brought her to the attention of top national Sierra Club officials. Thus she was well-positioned in the early 1950s to participate in the Club's first major foray outside of California: stopping the effort to build a dam in Dinosaur

National Monument on the boundary between Utah and Colorado and at the junction of two rivers, the Green and the Yampa. The dam would provide water storage and hydroelectric power for customers throughout the region. After federal agencies authorized the dam, the Sierra Club, Wilderness Society, Audubon Society, and other conservation groups pounced. Such a magnificent landscape had to be preserved. Additionally, conservationists believed that Dinosaur was just the first step in a larger campaign to privatize and develop national parks.[18]

Working with Brower, Jackson wrote letters to all 150 members of the Los Padres chapter and asked each to write letters to five other people. She provided names, addresses, and phone numbers of politicians and decision-makers in Washington. She also wrote articles about Dinosaur National Park in *Condor Call*: "Both the integrity of our National Park system and the saving of millions in tax dollars are involved." She put together a program in San Luis Obispo County and arranged for a room to show slides of Dinosaur, accompanied by music from Wagner's opera *Die Wulküre*. "We filled the hall!" she exulted. And she helped Brower and others promote float trips down the Green River, where conservationists mingled with journalists, photographers, and a documentary filmmaker. After a campaign lasting several years, the Department of the Interior pulled Dinosaur from its list of potential dam sites. Jackson realized early, she said, that getting media to pay attention to conservation causes was crucial to their success. This recognition would serve her well in her own campaign to save the Dunes.[19]

Her work on Dinosaur earned praise from Brower and elevated her profile even further. In 1956 the Sierra Club executive board tapped her to become chair of the council that oversaw all of the group's chapters, still mostly in California. The council administered "individual chapter concerns, disputes, territorial boundaries," she explained to one interviewer. At the time virtually no women held leadership positions in the national organization, though they were active in local chapters, often as newsletter editors. She served for two years and enjoyed the job, except for the meetings, which were "too long, just too much talk." That same year she moved with her family from Santa Barbara to Paso Robles, to be closer to Duncan's almond crop, which was sold each year to the Hershey Company for its candy bars. The family of seven lived in a five-bedroom Victorian home on three-quarters of an acre. Originally built in 1892, it needed a lot of work. The Jacksons remodeled the large kitchen, installed an intercom system, and built a pool with a cabana. Family activities were combined with Sierra Club obligations. Whenever the Jacksons took a trip, "I brought my typewriter," Kathleen said later.[20]

In addition to her Sierra Club work, she served as the women's coordinator for the campaign of Vernon Sturgeon, a Republican state senator from Paso Robles. "Get item in newspapers every time Vern sneezes," she wrote in a note to herself. She would later come to disagree with Sturgeon on the future of the Dunes. She also found time to go on long hikes with fellow Club members. Occasionally Duncan and their children accompanied her, but often she hiked without them. She did meet other men on the trail, including one named Paul, who wrote to her, "You are a remarkable woman Kathleen; however there's a bit of the divil in ye . . . and it's lucky I am to be liven no nearer to ye." Bob Thompson from Pasadena apparently wanted more than friendship, writing, "[After] just a few dozen sentences with you, I was an enthusiastic fan and ardent admirer."[21]

Duncan's money, professional stature, and political connections made Kathleen's activism possible because it enabled her to hire a housekeeper and to become acquainted with prominent politicians and businessmen. In reality, though, the couple had little in common. He was sedentary, focused on his music, his model trains, and his business interests. He was also much more conservative, both politically and socially. He was distinctly uninterested in her Sierra Club activities, though he admired some prominent male members, including David Brower and Bob Cutter from Cutter Laboratories, a family-owned pharmaceutical company.[22]

He appears to have been a willing host during Sierra Club gatherings at the family home. Attendees swam in the pool, sunbathed, and visited. At 6 p.m. the Jacksons fired up the barbecue; except for coffee and dessert, however, guests brought their own food and drinks. And Duncan seems to have been a sounding board for Kathleen's ideas and concerns, at least until he began to fear her activism was hurting his reputation. For the first few years in Paso Robles, Kathleen remained a member of the Los Padres chapter, even though meetings took place in Santa Barbara, more than one hundred miles to the south. In the early 1960s she was instrumental in starting the Santa Lucia group, based in San Luis Obispo. Still part of the Los Padres chapter, it was much closer to home—only thirty miles away. For someone who drove so much, it is perhaps not surprising that Jackson kept lists of mileage to and from various destinations: from home to the freeway, to San Luis Obispo, to the Nipomo underpass, to Highway 1.[23]

It was one of her children who introduced Jackson to the Nipomo Dunes for the first time. In July 1961 she promised her daughter Carol a special treat for her sixteenth birthday; Carol could pick out any place she wanted to go and Kathleen would drive her there. As she headed down the coast with a car full of teenagers, Kathleen was stumped as to their final destination. They passed San Luis Obispo, Pismo Beach, Arroyo Grande. Finally Carol directed her mother off the highway and down a long dirt road. They exited the car and trekked to the Dunes. Kathleen was moved beyond words, she later recalled. She left Carol and her friends sitting on the beach and began walking. The wind whipped her hair, the warm sand blew in her face. The vast and silent landscape wrapped itself around her. She began to cry, she recalled, not understanding why. "From that day on, she was obsessed" with the Dunes. Shortly after the first visit, she began hiking there regularly, sometimes alone and sometimes with Sierra Club members. On January 1, 1962, she led the first of what would become annual New Year's Sierra Club hikes.[24]

After one outing she stopped for lunch at a restaurant in the nearby town of Oceano. "It's so beautiful here," she said to the owner. "These dunes are magnificent.... Don't you think it would be great if [they] were preserved in a California State Park?" The owner looked at her for a moment, then replied, "It would ruin my business." It was a hint of things to come.[25] In fact the state Division of Beaches and Parks in 1959 had tagged the Dunes as a potential park. According to a Park Service survey at the time, "This large, unspoiled area possesses excellent seashore values and should be acquired for public recreation and conservation of its natural resources." Nothing further had occurred, however.[26]

"You can't take on PG&E," Duncan declared when she told him she planned to fight the proposed power plant. "I won't let you humiliate yourself." Brushing off his comments, she went to work. As the person responsible for garnering publicity for local Sierra Club outings, she knew editors at five local papers and reporters at radio stations. Through political work, and her husband, she knew San Luis Obispo County supervisors and other local and state officials. She put notices in newspapers, tacked them on billboards at Cal Poly, and contacted Sierra Club members, inviting them on a hike at the Dunes. Dozens of people showed up. When they reached the top of one dune, she stopped and began talking about PG&E's plans. She told those gathered about John Muir and his early twentieth-century fight to save Hetch Hetchy, in Yosemite Valley, from being dammed to provide water to San Francisco. "Once a resource is gone, it is gone forever," she told her captive audience. She assured those gathered that she was not opposed to nuclear power in general, but to the location of a power plant in such a unique area. She asked her fellow hikers to contact Ed Rayburn, a member of the Sierra Club executive board, and Brower.

Following the hike, she began an intensive lobbying campaign, personally visiting newspaper offices and attending meetings of the county board of supervisors and county planning commission, using the time allotted for public comments to make her pitch. She was most comfortable in hiking clothes, but at these gatherings she "was careful to dress as the socially respectable matron she was": two-piece suit with skirt and jacket, crisply tailored shirt, pumps, hat, gloves, and earrings. She exuded deference and never raised her voice. Over and over again she listened to the same refrains: "San Luis Obispo County needs the tax revenue from a power plant"; "What about the schools?"; "You're the only person in the region to oppose it." On many occasions she had to force herself not to cry in the face of patronizing condescension from male business representatives and public officials. She clenched her hands so audiences would not see her shaking with fear as she addressed all-male boards and clubs. But she kept going back.[27]

In mid-1963 Jackson was surprised by an invitation to a luncheon hosted by a group calling itself Friends of the Central Coast. It turned out to be a gathering of businessmen who strongly favored the proposed plant. Also invited were representatives of PG&E, who had heard of her campaign and wanted to see what they were up against. They introduced themselves and asked if she would be willing to meet. PG&E executives believed they held the upper hand—they were so confident, in fact, they thought they could breeze through the permit process. Initially they had characterized Jackson as "probably a rich busybody." But the Bodega Bay "debacle" had cost the company $4 million and had taught it to tread cautiously and to pay heed to opponents, no matter how insignificant and powerless they seemed to be.[28]

Jackson responded to the men with her characteristic approach, inviting them to a hike on the Dunes a few weeks later. Two PG&E representatives showed up. One was Kenneth Diercks, manager of government and public affairs for the utility. She had suggested they wear hats and clothing suitable for hiking and carry water. Instead they showed up hatless and wearing neatly pressed khakis and tennis shoes. They also failed to bring sunscreen or water. The trio trudged alongside agricultural fields, around Oso Flaco Lake, through brush, and up and down dunes. Jackson was nearly twenty years older than her fellow hikers, yet they struggled to keep up, slipping and sliding in the loose sand until they reached hard-packed sand at the water's edge. Fog hugged the ground as they began, then gave way to brilliant blue skies. When they stopped to rest, Jackson realized the men had not brought water or food. She shared a chocolate bar, but not her water.

At the end of the hike her guests were exhausted and sunburned, hungry and thirsty. They offered to take Jackson to a late lunch. She accepted, but insisted on paying her own way. "You may be the only person in San Luis Obispo County who really doesn't want PG&E to build our generation plant here," said Diercks as they ate. "So people keep telling me," she responded as she slowly sipped her soup. Then she asked a question whose significance would not be known for some time: "Isn't there some other place you can build?" Back at the San Francisco headquarters, Diercks warned his supervisors, "We've got a problem. . . . She's another Rose Gaffney. She's tough as nails."[29]

Jackson followed up by inviting Diercks to the Sierra Club's biennial Wilderness Conference in San Francisco, a gathering of America's most prominent conservationists. Diercks accepted. Like Jackson, he believed in building relationships with opponents. At the conference, she pointed out U.S. Supreme Court Justice William O. Douglas and Interior Secretary Stewart Udall. She introduced Diercks to Ansel Adams and David Brower. She also learned about a new environmental organization, Conservation

Associates, led by Dorothy Varian.[30] Jackson approached Varian, the widow of Russell Varian, founder of Varian Associates, an extraordinarily successful business that specialized in digital technology, linear accelerators, and other technological instruments. The Varians were part of a network that donated money to the American Civil Liberties Union, the Sierra Club, and other progressive groups.[31] After Jackson described her Dunes campaign, Varian told her, "We can't promise anything. Send us all the information you have, especially about geology."[32]

By the end of the evening, Jackson felt somewhat hopeful. But trouble soon emerged from an unexpected quarter: colleagues in the local Sierra Club. Not every conservationist, it turned out, opposed a nuclear power plant on the Nipomo Dunes.

Lee Wilson, like Jackson, was a pioneering member and leader of the Santa Lucia group of the Los Padres Chapter. Born in Colorado in 1905 and reared in Arizona and the small San Joaquin Valley town of Lindsay, he graduated from UCLA with a degree in electrical engineering. During World War II he taught at UCLA, and afterward he started an electrical contracting business. He joined the Sierra Club in 1948 and, with his wife, Lillian, led annual hikes each New Year's Day up Mt. Whitney. He first visited California's Central Coast when his son attended Cal Poly, and in 1958 moved his family, and his business, to San Luis Obispo County.[33] For the first few years of their acquaintance, Wilson and Jackson enjoyed a casual friendship. Early on, they alternated as leaders of the Santa Lucia group. He attended her Dune hikes, and she participated in hikes that he led into Lopez Canyon in the Santa Lucia Mountains east of San Luis Obispo. Wilson also had cordial relations with Duncan Jackson.

Disagreements seem to have arisen with Jackson's crusade to save the Dunes. They soon grew into disputes and arguments, and by 1964 tensions had flared into outright animosity. According to one account, the two were driving back from a hike in the canyons when Jackson began talking enthusiastically about her Dunes campaign. Wilson, who was driving, became red-faced; he finally exploded: "Why don't you just stay out of things. Stick to your knitting! You're making a lot of people mad." Finally he warned her, "I really don't think you're safe leading hikes out there."[34]

Jackson was stunned at the level of vitriol. She knew Wilson was trying to gain protection for Lopez Canyon, northeast of the Dunes, and pondered the possibility that he feared success for her might mean failure for his effort. But his hostility seemed more personal. Perhaps, as the knitting comment suggested, it was her gender. It was the early 1960s; the feminist movement was still a few years away, and as a woman she could be viewed as overstepping traditional boundaries, moving literally into men's spaces: boards of supervisors, Planning Department and City Council meetings, newspaper editors' offices. She had a brisk, take-charge manner. Alone, with no authorization from the Sierra Club, she had taken on the cause of the Dunes.

Jackson "wasn't bitchy," recalled the San Luis Obispo activist John Ashbaugh, who hiked the Dunes with her, "but she staked out her ground."[35] This approach might have worked in her personal life, but she was part of a large, national organization, led virtually entirely by men. She also was not shy about saying exactly what she thought, and she openly stated her belief that Wilson might be trying to sabotage her efforts because of his electrical contracting business. A nuclear power plant would need electrical components, so he had a self-interested motive, she implied. Whatever the cause, the rift only grew wider over time.[36]

Jackson could and did ignore Wilson, but she had to admit that his comment about "making a lot of people mad" was an accurate assessment. Virtually everyone in San Luis Obispo County supported the power plant: business owners, public officials, members of local chambers of commerce, and members of the Kiwanis, Lions, and Rotary clubs. In local stores, fellow shoppers purposely bumped into her. In the era before unlisted phone numbers, hang-up calls came into the Jackson home at all hours of the night. After one 2 a.m. call, Duncan became angry. "Is it smart to go against everybody?" he asked his wife. "People are after me all the time, asking if I can't make you stop." Among the most aggressive opponents were off-road-vehicle devotees, who spent weekends racing over the Dunes and feared Jackson's campaign might close the area to them. She tamped down her fears and kept hiking.[37]

She also continued to show up at public meetings. In early 1964 she attended a San Luis Obispo County supervisors' meeting. It featured PG&E's

Diercks, various Chamber of Commerce members, and pro-nuclear advocates in general. Speakers lined up to denigrate the Dunes as "a wasteland" and describe them as of "doubtful recreational and scenic value." When her turn came to speak, she started with her default opening, the damming of Hetch Hetchy, then moved on to the Dunes. "Industry does not belong on the wind-sculptured white sands and its green flowering glades," she said. She assured those in attendance that the Sierra Club did not oppose "power plants, piers, conveyors and factories": "We approve of industry in the proper location."[38]

As she finished speaking, Diercks approached her with an invitation to visit PG&E's small plant under construction at Humboldt Bay. She in turn invited Fred Eissler, a Santa Barbara teacher and a member of the Los Padres chapter and of the Sierra Club executive board, to accompany her. He declined. As it turned out, despite her assurances, some Sierra Club members were strongly opposed to nuclear power plants; they focused on issues such as the potential for radiation exposure, accidents, damage to ocean life, and earthquakes. Eissler was one such opponent. "I am personally concerned about health, safety and public welfare factors and a number of experts are too," he told her. When she told him about PG&E's invitation, he was dismissive. "You will be snowed," he warned.[39]

Duncan agreed to accompany Kathleen on the trip, though, as both partners recognized, their marriage was "no longer working." Their children were nearly grown, and within a year the couple would begin divorce proceedings. They were driven to San Francisco and then flown by private jet to the plant. Kathleen was not snowed. In fact she was troubled, though she said nothing at the time. When she asked a plant employee what PG&E planned to do with spent radioactive fuel, he replied, "We don't know yet." Diercks surprised her during the drive back to San Luis Obispo by asking where she would put a power plant, if she had a say in the matter. She talked about the many slot canyons along the coastline. There were oak trees, but otherwise the places were not unique, she told him. Upon her return, she phoned Eissler with the news. "I am going to do my best to get to know the people in PG&E and I want them to know Sierra Club people." He remained unimpressed, replying tersely, "You will fail."[40]

In early 1965 Jackson convinced the Sierra Club's president Will Siri to take a guided tour of the Dunes, the place he had heard so much about but had not yet seen. A biophysicist at the University of California's Lawrence Livermore Laboratory, Siri was an active member of Save the San Francisco Bay. And he was a world renowned climber, even leading an American expedition up the west ridge of Mt. Everest in 1963. He supported nuclear technology.[41] Before the hike, Jackson called members of the media—newspaper reporters and photographers, radio and television crews. Photos from January 16, 1965, show Siri, Jackson (referred to as Mrs. Duncan Jackson), and a group of hikers—including Jackson's critic Lee Wilson—clambering up dunes. The group "trudged, rather than walked about a mile-and-a-half" from Oso Flaco Lake to PG&E's land. At the end Siri quipped to reporters, "I sometimes thought it was easier hiking Mt. Everest."[42] He was awed by the dunes; he had not known how magnificent they were, he told Jackson, and he signed on to her campaign to save them. The backing of a leader of the nation's most prominent conservation organization—which then boasted nearly fifty thousand members—gave PG&E more incentive to find another location.[43]

Jackson believed that she had reason to celebrate, as she had been the driving force behind what might fairly be called a coup. She would offer input on the new site; after all, she knew the Central Coast as well as anyone, though final agreement on an alternative location would be the purview of the Sierra Club's national board of directors and PG&E. She was surprised, therefore, when Siri admonished her not to discuss any issues involving power plants to members of the media or the public without explicit authorization from the Club's national leadership. Siri also left her out of the public narrative describing how PG&E had come to agree to a new site, offering instead, "[It was] the result, I guess, of having said, 'we want the dunes preserved; go find another place.'"[44]

It is likely that Siri's admonition reflected the fact that he and Jackson, with input from Conservation Associates, had begun pushing for negotiations with PG&E without authorization from the board, which was against protocol. Nonetheless it meant Jackson was left to deal with any local fallout, though she could not personally respond or fight back. And fallout quickly ensued. State Senator Vern Sturgeon, whom Jackson had long supported, opposed a new location for the proposed plant. Fewer than 1 percent of San Luis Obispo County residents favored preserving the Dunes, he declared. Why should this tiny contingent prevail over the vast majority that opposed preservation?

Dozens of county residents penned letters to the editors of local papers citing the same issues opponents had always raised: loss of tax revenue, loss of jobs, schools left with less money. Jackson's old nemesis Wilson moved back into the frame, seeking to oust her from the Santa Lucia group's leadership. According to Jackson, he wrote letters to some members suggesting that she had something to gain personally from preservation. Perplexed and irritated, she wrote to Siri: "I do not know what he can do to harm me personally, but he might possibly cause a public stir that would be very bad public relations for the Sierra Club."[45] Jackson continued her work, though she had to convince other people to write letters and make phone calls on her behalf. She also continued to publicize hikes in the chapter newsletter. In March 1965 the headline "Can You Help Save the Nipomo Dunes?" spanned the entire front page. The story below began, "When the sun is bright over the sculptured white sands ... the sea blue intensifies."[46]

Neither she nor anyone else could have guessed that the Dunes campaign was about to unleash a "civil war"—or "uncivil war," as some called it—in the Sierra Club, one with far-reaching ramifications.[47] Club leaders had developed strong and divergent opinions on nuclear energy in the years since PG&E first announced its intention to build a plant on Bodega Head. One faction saw the inevitability of nuclear power and believed that compromising with utilities and government agencies might win them some concessions on safety and siting. A competing group was appalled by any potential compromise with PG&E; it believed nuclear power "represented a catastrophic threat to society."[48]

The emerging divisions reflected challenges facing long established groups, which were being pushed to expand their agendas in order to remain relevant. For decades the Sierra Club had attracted mostly white middle-aged professionals who loved the wilderness and wanted to preserve it for recreational use. By the mid-1960s, however, more people across the country were signing on to a new, broader environmental movement, partly as the result of proliferating grassroots campaigns fueled by citizen activists, many of whom were women. As a result, older, male leaders of established groups were forced to confront sometimes unwelcome changes. Jackson reflected the underlying tensions. She was a faithful Sierra Club member and local leader, but her passion for and obsession with saving the Dunes placed her in the citizen activist camp as well.

The possibility of a civil war within the Sierra Club was not apparent when PG&E, working with Jackson, a few other Club members, Conservation Associates, and a state resources agency task force, began searching for an alternative location. Ultimately they settled on Diablo Canyon, near the small town of Avila Beach. From a distance it seemed ordinary—a "slot canyon," in Jackson's words, similar to dozens of other grassy, oak-strewn valleys abutting cliffs that dropped off to rocky outcroppings and beaches along the coastline. Jackson briefly visited Diablo and believed that Club leaders would approve the choice. In May 1966 the Club's board of directors gathered in San Francisco to discuss Diablo. It was the first time the board as a whole had learned of the talks between their members and PG&E. In her testimony Jackson described Diablo as "a treeless slot" in an "isolated coastal canyon." Any power plant could be seen only from the water, she added.[49]

Siri told the board that approving Diablo was the only way to save the Dunes. Voting 9 to 1, with two abstentions, the board determined that "the Nipomo Dunes should be preserved, unimpaired for scenic and recreational use under state Management.... Diablo Canyon ... was a satisfactory alternative site." Fred Eissler, the lone dissenter, was outraged. Recommending Diablo, he asserted, "was tantamount to approving a dangerous nuclear generator." He refused to back down. Four board members had been absent at the time of the vote, and Eissler sought their help to overturn it. Martin Litton, a newcomer to the board, agreed with Eissler, but for a different reason. An avid photographer and the travel editor at *Sunset Magazine*, he actually had seen and photographed Diablo and took strong exception to its description as a "treeless slot." It was, instead, "*the* representative area of the whole California coast," with "the largest standing coast live oak." It was also "the best preserved tidal zone in California," containing "masses of abalone and sea lions." And it contained a sacred burial ground for the Chumash tribe. Litton showed his photos to Brower, convincing the executive director that the vote supporting Diablo had been a "gross mistake."[50]

Jackson's own local Sierra Club chapter also had been angered by the choice of Diablo. They had been blindsided by the news, according to one member: "A good many of the local group … [who were not notified in any way] were violently opposed."[51] Jackson was "pilloried," as she later termed it, for misrepresenting Diablo. But she was angry with herself as well. She had to admit she had just walked to the mouth of the canyon and no further. "We had not seen oak trees, nor woodland, only closely cropped, closely graded hillside." Stopping short had been a mistake: "I felt very disturbed. I wrestled with my conscience." She understood that "it is a grave responsibility … to alter forever a piece of God's earth." But she remained committed to the Dunes: "I felt the saving of the Dunes … a more unique piece of the earth than Diablo Canyon, was the place of primary concern."[52]

She was further embarrassed when one Sierra Club official called her out for comments she made after PG&E announced it had leased six hundred acres in Diablo Canyon from the rancher who owned the land. Jackson had called it "a great day." George Marshall said he was "shocked" by her comment. He viewed it, instead, as a "sad day." He further remarked, "It is not the function of the Sierra Club or any of its representatives to be apologists, public relations people, or anything of this kind for PG&E or any other power company nor for nuclear power."[53]

Opponents of Diablo pushed the executive board to revisit the issue in fall 1966, seeking changes in the original agreement, plus a moratorium on building nuclear power plants altogether. A second vote upheld the initial decision, approving the moratorium but without the inclusion of Diablo. By this point Brower had become, as he put it, "a born-again anti-nuclearist." He agreed with Litton and Eissler that California did not need nuclear power plants.[54] Siri empathized to some degree: "As a conservationist, I have to agree that no development along the coast would be desirable, but we live an an energy-based society." He bemoaned the conflict, which was sapping "the club's energy." The anti-Diablo side refused to give in, calling for a referendum in which all Sierra Club members could vote on the issue. In its seventy-five-year history the Club had never held a referendum.[55]

Acrimony spilled into public view, as the media began covering the internal conflict. Gladwin Hill of the *New York Times* dubbed it "a showdown between devotees of the organization's original role as a 'hiking club' and those favoring its new role as a watchdog against despoliation of the natural environment."[56] Jackson called it a battle between "appeasers" and "purists," counting herself in the first category. Some Sierra Club members blamed her for their problems, she realized, since she had started it with her Dunes crusade.[57]

By a 2-to-1 margin, the April 1967 referendum upheld the board's original decision on Diablo, but the vote did not end the controversy. There were arguments about fairness with regard to published positions regarding the referendum; some Diablo opponents charged that Ansel Adams's stunning black-and-white photos that accompanied the pro-Diablo side's argument had given it an unfair advantage. There were more board votes, each one upholding the original decision, though by a narrower margin than the one before. In September 1968, though it changed nothing, the board acknowledged by a 9 to 5 vote that it had "made a mistake of principle and policy in attempting to bargain away an area of unique scenic beauty."[58]

Opponents of Diablo tried different strategies, suggesting that PG&E choose another site, possibly inland, utilizing another water source, or at Moss Landing in Monterey County. Some members of the Club's Santa Lucia group who opposed the Diablo location created their own organization, the Scenic Shoreline Preservation Conference. Its mission: to closely follow developments and appeal approval of PG&E actions to state and federal agencies. In 1967 Scenic Shoreline appealed a California Public Utilities Commission decision to approve Diablo, charging that the Commission had "pitted the resources of the largest public utility of California against ... members of the public who [did] not command the funds and expertise necessary to conduct independent investigations." The group lost the appeal; the California Supreme Court denied Scenic Shoreline's request to stop construction. Despite these setbacks, Scenic Shoreline continued to battle the utility, focusing largely on seismic concerns.[59]

By 1968 Brower had become the focus for both sides in the conflict. Diablo was the immediate catalyst, though other issues—mostly having to do with the way he spent money—had long concerned some Sierra Club board members. Brower served at the pleasure of the fifteen-member board, which had hired him as executive director in 1952. Under his leadership, the Club had gained international prominence and prestige. But some board members believed that this prestige came at a high price. In the 1950s Brower had inaugurated a program to publish oversize books featuring glossy nature photographs accompanied by text. All of the books won high praise from reviewers and sold well—not well enough, however, to offset the cost. Brower's intensive focus on the books program raised his profile but meant less time spent on other Club concerns. From Brower's perspective, the board's hand-wringing over every penny meant he sometimes had to act without its approval, leading to charges that he was "insubordinate." Some board members also reeled at Brower's 1966 decision to submit a full-page advertisement to the *New York Times* opposing a proposed dam in the Grand Canyon. In response, the Internal Revenue Service revoked the Sierra Club's tax exempt status.[60]

But it was his pivot on nuclear power, specifically his effort in 1968 to pack the board with like-minded directors, that ultimately fueled Brower's downfall. Board members served three-year terms, staggered so that five members ran each year for election. In 1968 Brower put together a slate of candidates who opposed Diablo, and nuclear power in general. They called themselves "ABC—aggressive, bold, constructive." A slate of moderates, dubbed "Concerned Members for Conservation" was led by Brower's longtime friend and now antagonist, Ansel Adams. The Brower slate prevailed, meaning that the majority now supported his anti-Diablo position. His victory was not complete since the the board did not fully reverse its original vote on the power plant siting, but it ensured that the following year's election would be all about him. In fact Brower decided to run for the board himself in 1969, leading his old friend Wallace Stegner to declare that he had been "bitten by the worm of power." Brower took a two-month leave of absence from his executive director duties to prepare for the election.[61]

Both sides girded for an epic battle, which played out "largely within the pages of chapter newsletters."[62] The Santa Lucia group was no longer affiliated with the Los Padres chapter, but was its own separate chapter. It was not immune to the divisions wracking the Club as a whole. Not surprisingly, Jackson and her nemesis Wilson stood on opposite sides of the divide. Jackson supported the moderates: "Five men whose demonstrated stability, sober responsibility and *integrity* can help us find the way are five to vote for: Ansel Adams, August Fruge, Maynard Munger, Judge Ray Sherwin and Edgar Wayburn." She later described Brower as "gifted with words" but someone who "seemed to have an idee fixe from which he could not be swayed."[63] She might have been talking about herself.

Wilson, still trying to garner support to preserve Lopez Canyon, stood strongly behind Brower, who, he stated, "will lead us into a great tomorrow." Wilson's wife, Lillian, seconded his remarks, pointing out, "[Brower] is known far and wide as Mr. Sierra Club.... [Over many years] I have always been impressed by the responsible manner in which he has done his job." Some local members were exhausted with the battle and just wanted it to be over. Santa Lucia's newsletter editor Eileen Kangel said, "We feel the issue [of Diablo] shouldn't be placed on the ballot again. The Board of Directors made a mistake *then* by not investigating thoroughly all aspects of the situation locally."[64]

In the end, Brower lost the election, as did other candidates on his slate; the moderates had prevailed. He submitted his resignation, which the board accepted. The Dunes would be saved and Diablo would be the site of a nuclear plant. The Sierra Club board continued to support nuclear power into the 1970s, though some members voiced concerns, including fears that "a loss of coolant accident could cause death and injury from radioactive poisoning."[65] The discovery of an earthquake fault in the ocean near Diablo eventually turned the Club against nuclear power altogether. But its opposition, and that of other groups, failed to stop construction. At the meeting following Brower's resignation, board members began to discuss a broader agenda, encompassing "issues of open space, water quality, power developments, and radioactivity."[66] Meanwhile Brower founded a new organization, Friends of the Earth. Headquartered in San Francisco, it "quickly became a leader in the international antinuclear effort," advocating for a moratorium on nuclear power "until the environmental and genetic safety of their operation can be proved." Brower also became involved in the zero-population growth movement, which posited that too many people on Earth meant more environmental problems, an argument that many people saw as racist and classist.[67]

While the Sierra Club was mired in the Diablo controversy, Kathleen Goddard finalized her divorce, took back her maiden name, and moved to the town of Arroyo Grande in southern San Luis Obispo County. Her primary focus had always been the Dunes, and now she spent much of her time trying to convince local and state authorities to set aside money to turn them into a park. Not surprisingly, she ran into roadblocks. She discovered that Collier, a chemical corporation and subsidiary of Union Oil, planned to build a conveyor to carry the byproduct of Union's refined oil from rail cars across the dunes to a wharf, where it would be loaded onto ships. Goddard responded the way she always did, by organizing hikes; she called them "Sundays in the Dunes." Hundreds showed up. The conveyor idea was abandoned, but mostly because of opposition from Southern Pacific Railroad. A sand mine had also been proposed, but that idea was abandoned as well, as was a proposal by a developer to build homes near the Dunes.

In 1964 California voters had passed Proposition 1, setting aside $150 million for parks throughout the state. Goddard labored to convince the San Luis Obispo Board of Supervisors to include the Dunes on its county list for park funds. She was aided in this effort by leaders of Conservation Associates, who had visited a geology professor in Boulder, Colorado. He called the Dunes "unique" and like "no other formation ... in the world." Goddard was disappointed, however, to discover that the supervisors considered the Dunes a low priority: eleventh for land to be added to the Pismo Beach State Park, and twelfth for acquisition of Oso Flaco Lake. The California State Parks Commission ignored the county's recommendation, however, placing the Dunes seventh on its statewide list. The proposed park "would include 32,000 feet of ocean frontage" as well as the "freshwater dune lakes." Business interests tried to get the Dunes removed from the list, but the Parks Commission ignored the request. At an estimated cost of $7.5 million, however, it would take some time to accomplish, for the Commission had approved "no purchases of more than $4 million."[68]

In her usual fashion, Goddard showed up at Parks Commission meetings to continue pressing for park status. At one meeting she told the commissioners,

"The dunes offer a rare type of scenery, unique in the west and perhaps in the country, entirely different from other dune complexes in the United States. Here is a beauty of surpassing serenity, a place of stillness, to save our sanity in a troubled world." Commissioners provided assurances that the Dunes were "still a 'live project' for park acquisition," though they proffered no timeline.[69]

By the late 1960s Kathleen Goddard was exhausted. She felt that she had to correct every misstatement and erroneous comment made about the state and its desire, or not, to acquire the Dunes. In March 1969 she wrote a letter, and a correction, to be published in all Sierra Club chapter newsletters rebutting "flagrant untruths," specifically suggestions that many parts of the Dunes had already been saved, and thus the state had no need to acquire them: "The fact is that thousands of unspoiled, natural, beautiful, open dunes and rugged cliffs wait our 'saving effort....'. Moreover, the State of California has shown acquisition interest. Purchase awaits negotiation with owners and four to seven million dollars from a new State Park bond issue." But threats remained, including sand mining and an extensive subdivision proposal.[70]

She often felt on the verge of tears and experienced frequent shortness of breath. She also needed to find a paying job—her divorce had left her low on money—but she had difficulty obtaining work. She no longer had her husband's connections, and she had alienated many people. Potential jobs emerged and then disappeared. One job with a lawyer ended abruptly after only a few weeks. "It began to dawn on her that she had paid a very high price for the Dunes."[71] Likewise, some in San Luis Obispo County feared that Diablo might never be completed, costing the region millions of dollars with no return.

In 1970, at the age of sixty-three, she decided that to clear her head and restore her equilibrium, she needed to get away. Alone, she embarked on a pack trip through the high Sierras. She wrote "to the head ranger at the Inyo National Forest, explaining she would be exploring the area alone, enclosing an itinerary. She asked that the rangers keep an eye on her. In return, she pledged to fill plastic bags with litter wherever she went." A friend dropped her off at the trailhead. She struggled to carry her heavy pack until a family picked her up and took her to her first destination. The next day they returned to take her to Kern Lake. There she obtained a pack burro, which became her companion for the remainder of the seven-week trip. Other backpackers along the trail "brought home tales of a strange little woman with a burro who divided her attention between picking up cigarette butts and identifying wildflowers."[72]

She returned home reinvigorated. Despite some strained relationships, she was still active in the Sierra Club. Lee Wilson was now chair of the Santa Lucia chapter. He led hikes in Lopez Canyon, where he was still seeking wilderness status. Meanwhile, Goddard led hikes in the Dunes, including evening hikes. She attended botany classes at Cal Poly and learned the Latin names of plants. She joined several other environmental organizations, including the Wilderness Society and the Audubon Society, and decided to create a local chapter of the California Native Plant Society.

At the Native Plant Society's first annual dinner, she met Gaylord Jones, a widower, retired architect, and avid nature photographer. They married in May 1971, honeymooned in the high Sierras, and then settled in a small home in Arroyo Grande, "delightfully cluttered" with Dunes memorabilia. One wall held framed and autographed photos by Ansel Adams. They regularly hiked the Dunes, "dressed alike in plaid yellow shirts ... with broad-brimmed hats and sturdy walking sticks," devoted to learning the names of all the plants.[73] On one hike they came across a two-room cabin and learned that it belonged to Bouke Schievink, the last surviving member of the fabled Dunites. They introduced themselves and thereafter visited whenever they could, though they soon discovered that Schievink could tolerate only short visits.[74] "Gaylord was sweet," recalled John Ashbaugh. He was content to remain in the background and "give the major platform to his wife."[75]

As she neared seventy, Kathleen Goddard Jones embraced a new generation of activists. She and Gaylord attended an Ocean and Conservation Conference in Los Angeles. The gathering, she wrote, included "surfers, fishermen, sunbathers, beachcombers, oldsters, sailors, concerned citizens' groups and *farsighted* members of the business community."[76] She visited local schools, talking to the students about environmental issues, and occasionally led students to the Dunes. As Ashbaugh recalled, "She made everyone feel so welcome. She invited the kids to roll down the face of Dunes with her."

The kids crossed their arms, grabbed their opposite shoulders, and hurled themselves off the top of the Dunes. When she ran into the students while shopping, they would exclaim, "That's the dune lady. She came to our school and talked about birds and flowers. It was cool."[77]

In 1974 she received word that PG&E planned to sell 857 acres of the Dunes, along with Bodega Head, to the California Parks Department. But it was not a complete victory. Part of the Dune land would be set aside for off-road vehicles. She sought out the ranger in charge of the section and eventually worked with him to close most of the Dunes to vehicles and fence off the portion where off-road vehicle enthusiasts could ride. In 1980 the U.S. Fish and Wildlife Service issued a report, "California's Important Fish and Wildlife Habitats," which listed the Dunes—"among the highest aesthetic and ecological values remaining in California"—at the top of forty-nine sites. The agency recommended that between three thousand and ten thousand acres be preserved. That same year, the federal Department of the Interior "designated the Nipomo Dunes as a National Natural Landmark."[78]

As she worked to make the park proposal a reality, Goddard Jones continued her weekly hikes, forging "briskly ahead down into the canyon headed straight west," wrote one admirer. In 1979 she was hospitalized for an undisclosed ailment. "Without leaders like Kathy," a fan wrote, "the Sierra Club could barely function."[79] When the state finally purchased a portion of the Dunes in 1983, she attended the ceremony. At seventy-five she was no longer particularly active in the Sierra Club, except for matters involving the Dunes. But a new generation of San Luis Obispo activists had emerged. They had not been party to the acrimony involving her role in selecting Diablo Canyon, and many viewed her as a role model for how to pursue environmental goals.

In the mid-1980s the Sierra Club embarked on an oral history project, documenting environmental causes and prominent people who had worked on them through the decades. Ashbaugh, then a young San Luis Obispo activist, wrote one of two introductions to Goddard Jones's entry: "Kathy is the type of volunteer every Sierra Club chapter should have: resolved, knowledgeable and relentless. Her battle on behalf of the Nipomo Dunes represents the finest qualities of courage and commitment.... The history of the Sierra Club, particularly on the Central Coast, could not be written without devoting a major chapter to . . . our redoubtable lady of the Dunes. Kathy has led hikes beyond measure, attended countless meetings . . . and written literally mountains of correspondence."[80] Dirk Walters wrote the other introduction. He had been a young botany professor at Cal Poly when he joined the Native Plant Society. His first impression of his fellow members: "How old they were." But he soon changed his mind: "I have looked on in amazement while Kathy has mobilized the public [for the Dunes]."[81]

In 1982 a researcher contacted her for comments on Diablo Canyon and on the divisions that had enveloped the Sierra Club nearly two decades earlier. "Why exhume these painful memories?" she asked him. Though many people had blamed her for the debacle, in reality she hadn't had much input into the choice of Diablo, she insisted. "I didn't really have that much knowledge about it." Memories of that time still clearly haunted her, however. She acknowledged to one interviewer that she kept a folder full of letters from Sierra Club members who had criticized her—using wrong information that needed to be corrected—and complimented her.[82]

Gaylord Jones died in 1991. Then in her eighties, Goddard Jones had finally begun to slow down, but mostly out of necessity. She was hard of hearing; she broke a hip, but three months later exulted, "As of today . . . at last I am allowed to drive. (Don't ever break your hip, especially if you are over 80.)"[83] Walking home on Thanksgiving night in 1995, she stepped off a curb and into the path of a truck. She was seriously injured but eventually recuperated. In July 1997, on her ninetieth birthday, she led a short hike in the Dunes. By this point she had garnered recognition far beyond California's Central Coast. In the early 1990s the state Coastal Conservancy publicly recognized her for all of her work on the Dunes.

That same decade saw the creation of an alliance among local, state, and federal agencies, businesses, and private landowners to preserve the Dunes. They included San Luis Obispo and Santa Barbara counties, the Land Conservancy of San Luis Obispo County, the California State Parks Department, U.S. Fish and Wildlife Service, Vandenberg Air Force Base, farmers, and oil companies. Negotiations were led by the Nature Conservancy, an environmental organization dedicated to purchasing and preserving natural resources throughout the United States. Regional Director Steve McCormick dubbed the eventual agreement "an example of how environmentalists and their historical enemies can work together." But environmentalists did not get everything they wanted; 1,500 fenced-off acres on the beach in Oceano—under the jurisdiction of the California State Parks Department—remained set aside for off-road vehicles. On the other hand, nearly 2,600 acres purchased earlier by the Nature Conservancy were turned over to U.S. Fish and Wildlife Service for what became the Guadalupe Nipomo Dunes National Wildlife Refuge. It is home to many endangered species, including pelicans, peregrine falcons, snowy plovers, and least terns. This was added to thousands of acres already preserved.[84]

In June 2001 a reporter for the San Luis Obispo Tribune wrote a tribute to Goddard Jones that might have been her epitaph: "Before she emerged victorious in 1974, Jones was divorced and unemployed, had sparked major rifts in the local and national Sierra Club leadership and lost a fight to ban vehicles from what is now the Oceano Dunes State Vehicular Recreation Area." But she had also facilitated the protection of eighteen miles and eighteen thousand acres of coastal dunes. Goddard Jones suffered a stroke in late summer 2001 that left her partially paralyzed. She died in October 2001 at the age of ninety-four.[85]

Today a visitor to the Dunes can stop first at the Dunes Center, a small frame house in the town of Guadalupe that contains maps and geological information, as well as details about DeMille's Ten Commandments—in late 2017 archaeologists unearthed a three-hundred-pound plaster sphinx from the film, with most of its original paint intact. The Center also has a section on the Dunites, and it features a large portrait of Kathleen Goddard Jones looking toward the horizon, wearing her hiking clothes—hat, trousers, shirt—and holding a walking stick. On an adjacent table sits the battered green Hermes typewriter on which she typed out hundreds, if not thousands, of letters, press releases, newsletters, and notes to herself over nearly four decades.

After a lengthy and frustrating battle, Lee Wilson finally achieved wilderness status for Lopez Canyon in the Santa Lucia Mountains. Unit 1 of the Diablo Canyon nuclear power plant began commercial operation in 1985; unit 2 followed a year later. Back in the 1960s PG&E had estimated the cost at slightly more than $300 million. The final tab: more than $5 billion. By the late 1970s the drumbeat of stories about potential disasters at Diablo and the actual partial meltdown of Three-Mile Island nuclear power plant in Pennsylvania had turned a large number of Californians against nuclear power.

Among the most vigilant and vociferous were members of Mothers for Peace, representative of a new, more confrontational generation of activists. The San Luis Obispo–based group began in the 1960s by protesting the Vietnam War and then morphed into anti-Diablo activists. They called themselves the Abalone Alliance for the thousands of abalone and other species killed by warm water from nuclear reactors. Members held vigils, organized protests, and blockaded the entrance to Diablo Canyon.

PG&E and other utilities once envisioned dozens of power plants throughout California. Only a handful were built; all but Diablo have now been shut down, and it too is scheduled to be shuttered, in 2025. California will have no more nuclear power plants. It does, however, have the Nipomo Dunes, the largest intact coastal dune ecosystem on Earth.[86]

4

Fig. 4. Developers descended on Los Angeles after World War II. Mountain residents Jill Swift, Susan Nelson, and Margot Feuer (*left to right*) led a lengthy and ultimately successful effort to preserve mountain land for a vast urban park system. Photo courtesy of UCLA Special Collections.

SAVING THE SANTA MONICA MOUNTAINS

On Harriett and John Weaver's first date, in 1936, he took her to the roof of his apartment building in Kansas City, Missouri. She told him, "Someday I would like to live on the highest hill I can find." She was twenty-three and he was twenty-four. He was an editor at the *Kansas City Star* newspaper; she was a Phi Beta Kappa journalism graduate from the University of Kansas, and he had offered her a job reviewing books. Little more than a decade later, the couple, now married, lived in the Santa Monica Mountains overlooking the famed Sunset Strip in Los Angeles. The Weavers bought their $25,000 home in the late 1940s after John sold a screenplay that became the 1949 romantic film *A Holiday Affair*. Extremely frugal and abhorring debt, Harriett told the stunned realtor that they did not need a mortgage. "Oh, we always pay cash for things," she said.[1]

Early in their relationship, John discovered that Harriett was the handiest person he had ever met. Once, when a venetian blind broke, she picked it up and "after threading it with fresh white cord, rehung it." On a vacation to visit John's family, Harriett got stuck in the guest bedroom. The door was jammed, so she took it off its hinges. "I thought I'd married an Edison," he recalled. Her skills came in handy when they bought the hilltop house, since it had many problems. On the outside, it resembled a "box of cement brick." Inside, the living room featured wood beams covered in so much shellac they had turned orange over the years. It was dominated by a huge brick fireplace covered with a grill. "The two bedrooms were done in the same spectacularly bad taste." But the couple had bought it for the view, so Harriett quickly went to work. She rebuilt the fireplace, scraped off shellac, and caulked the windows. When John brought home a television set, she built a cabinet to hold it. Outside, she dug, moved dirt, and planted flowers, bushes, and trees on the hillside. She even built a fence. Her "day always begins with the drawing up of a list of things to be done," John wrote in a memoir that essentially amounted to a paean to his wife.[2]

Over time, others took note of her talents. In December 1953 the *Los Angeles Times* featured the Weavers and their home, which at the time of purchase had been "unique, in the manner of a two-headed cow." But there was that "superlative view of the city in one direction and of the hills in the other, a fairly wide area of flat land beyond the house for extension purposes and an existing house of 1200 square feet that had possibilities." Photographs accompanied the story. One depicted the house "perched on its hill" with a wide valley falling away in the distance, dotted with tiny roads and the city far below. Several other photos illustrated Harriett's handiwork in the kitchen, living room, study, and bedroom. With an acacia tree outside the master bedroom casting scattered sunshine across a wall, the couple, it seemed, practically lived in paradise. For a while, anyway. A decade or so after her arrival, Harriett Weaver would become one of dozens of women who labored to save the Santa Monica Mountains from developers.[3]

The Weavers moved to Los Angeles at a time when the city, and Southern California in general, experienced exponential growth. From a town of fifty thousand in 1890, sixty years later Los Angeles held nearly two million residents and the surrounding areas hundreds of thousands more. In the early twentieth century, people arrived from across the country to work in the oil, automobile, and aviation industries and in Hollywood. During World War II, Los Angeles became a center for the defense industry, which only grew larger in the early years of the Cold War. The pressure from war-time and postwar workers created a serious housing shortage, alleviated only by the mass production of standardized tract homes in the 1950s and 1960s. The city had long featured cycles of boom and bust courtesy of real estate speculators, the *Los Angeles Times*, and the Chamber of Commerce. But the postwar growth was unprecedented. Between 1950 and 1960 the population grew by more than 500,000.

Los Angeles was not the only place that saw massive growth during this period, but its pattern of development differed significantly from that of its counterparts elsewhere. The historian and writer Carey McWilliams describes Southern California in the 1940s as "a vast expanse of land unbroken by rivers, mountains, or other barriers.... With real estate companies fostering the centrifugal movement, Los Angeles became a city without a center." In the 1910s the city covered slightly more than a hundred square miles. That figure had jumped to nearly five hundred square miles by 1930, and to nearly six hundred square miles by the 1950s. Much of the growth was achieved by annexation of surrounding communities, nearly fifty in all, a phenomenon made possible because of the city's control of the region's water supply.[4]

By midcentury Los Angeles had spread north into the San Fernando Valley and south toward the harbor at San Pedro. Towns and cities that seemed to have their own discrete identities were, in reality, part of Los Angeles. They included Eagle Rock, Wilmington, and Northridge and many communities in the Santa Monica Mountains, "a heroic spine . . . that stretches from the ocean at Point Mugu across 200,000 acres to the playgrounds at Griffith Park."[5] Dipping into Ventura County to the north and west, the mountains are "a 47-mile collection of peaks and canyons offering wild relief from the solid geometry of a city." Mountain communities inside Los Angeles city limits include Brentwood, Bel Air, Sherman Oaks, Tarzana, and Woodland Hills. One of the mountain towns in Ventura County is Thousand Oaks. By the 1960s nearly eight million people lived within an hour of some portion of the Santa Monicas, the only mountain range in the United States "to bisect a major American city."[6]

When the Weavers moved to the mountains, the area was "a semi-wilderness with few roads or homes but many deer, raccoons and coyotes." At war's end fewer than thirty thousand homes dotted the hills, which contained

Topanga, Benedict, Coldwater, Laurel, and more than a dozen other canyons. Many celebrities called the area home, but not all of the neighborhoods were pricey. Beverly Glen Canyon, for example, was relatively inexpensive, populated by a diverse array of residents, such as teachers, musicians, and artists. Nine miles across at their widest point, the mountains rise from sea level to three thousand feet. The eastern side was and remains much more heavily populated than the western, which features rugged terrain containing manzanita, scrub oaks, sycamores, a kind of holly called toyon, and laurel sumac, as well as the oceanfront communities of Malibu and Zuma Beach.[7]

Developers and real estate investors had eyed the hills for homebuilding as early as the 1920s. Because lack of access to building sites hampered construction, city leaders promoted a $1 million bond measure to build a road that would run from Highway 101 in Hollywood, meander for more than twenty miles across the ridge of the Santa Monica Mountains, then drop down to Woodland Hills. Mulholland Drive was completed by the end of 1924.[8] The road made the mountains accessible to cars, but getting construction equipment to building sites remained daunting, and prohibitively expensive. Banks refused to lend money on hillside homes because they tended to be "original in design, picturesque, eccentric, but not sure commercially." As long as cheap land remained available in the vast Los Angeles basin, developers had little incentive to build in the hills, except for lower elevations such as Bel Air.[9]

The post–World War II period changed everything. Builders could not keep up with the demand—and not only for single-family homes. Lower-income workers needed cheaper housing, so builders began erecting condominiums and apartments, significantly altering the urban landscape. By the mid-1950s, as developers began to run out of available land in the Los Angeles basin, they eyed the hillsides and mountains. "With Los Angeles gaining half a million persons every decade . . . these steep, brushy, flammable mountain slopes" proved irresistible. New kinds of construction equipment—bulldozers, skip loaders, and larger trucks—made it easier to access and dig into steep and rocky terrain. And newly built freeways made it easier to get to building sites. Heavy equipment began arriving in the hills, reducing "primeval green slopes" to "crumbly, desolate cliffs, ripping at ancient seepages and springs, dumping topsoil in creek bottoms." The contractors "gouged out little shelflike 'pads' for bungalow or castles" and "shoved tons of loose dirt and brush over the side, where it slouched, loosely held up by bushes," wrote John Weaver.[10]

Most Southland residents initially supported developers because they built desperately needed homes. But by the late 1950s and early 1960s some began to question the consequences of unbridled growth, including the loss of natural habitat and an increase in multifamily housing units in single-family neighborhoods. While addressing the needs of lower-income residents, these units threatened the ambience of upscale neighborhoods and potentially lowered property values. Homeowners' associations began proliferating across the Southland, lobbying public officials to establish and then promote rigorous guidelines on building and zoning.[11]

One of the most powerful was the Federation of Hillside and Canyon Associations, created in 1952. It was composed of residents throughout the Santa Monica Mountains, among them John and Harriett Weaver. Individual mountain neighborhoods had their own associations as well. The catalyst for the Federation and its subsidiaries was a January 1952 storm that dropped more than seven inches of rain in two days—half the average annual total for the region. In the mountains, as one Beverly Glen resident, Richard Lillard, recalled, "tons of rainwater soaked into the tons of loose dirt and rock in the raw cuts and loose fills, until finally the soggy masses uprooted the shrubs they lay on and slipped rumbling and smashing down the slopes. Rivers of rain on denuded hillsides picked up rocks and soil and roared down." At least eight people died, and damages ran into the millions. Affected residents were cut off from the rest of the city and region since many roads "were closed to general use."[12]

Homeowners living directly downhill from new construction sites experienced the most serious damage, with thick mud covering their property several feet deep and, in some cases, burying their homes. It took weeks to dig out. Residents blamed developers, but also city and county officials who allowed the construction and who now seemed dazed by all the damage. Mountain residents descended on Los Angeles City Hall, demanding action. Their prominence, money, and connections enabled them to hire geologists and other experts and to enlist the help of celebrity residents, such as the film director Howard Hawks.[13] By the end of 1952 the Federation had notched its first victory. The City Council set up regulations for grading and filling in the hills, the first such ordinance in the country. All cuts needed to have "1.5 feet of horizontal distance for every 1 foot of vertical distance," later increased to two feet for every one foot.[14]

The victory was temporary, however. Grading rules failed to deter developers, who had the support of many politicians in the region. By the late 1950s they were adding whole subdivisions to their plans, as well as condominiums and apartments. Eyeing tax revenues, city planners said they envisioned as many as 200,000 residents in the area, a plan that enraged and terrified members of the hillside federations.[15]

One day in 1960 Harriett Weaver looked out her kitchen window to see bulldozers lined up across the street. As she later described it, "Neighbors began to gather in small, horrified groups, peering up at the yellow monsters systematically gnawing away at the helpless land at the top of our hill, worrying it into the succession of handkerchief-sized lots which was to pass as a subdivision, and spitting the dust back down into our faces." The city had given residents no advance warning since the planned subdivision required no zoning changes. This was not acceptable to Weaver and her neighbors, who protested the planned "40 to 50 houses, stair-stepped and cantilevered above one another on the ridge of the hill [that] would not only depreciate property values in the established area below it, but would also create fire and traffic hazards for the entire neighborhood."[16]

Once again mountain residents descended on Los Angeles City Hall, described by Weaver as "a great, faceless bureaucracy of boards and commissions, all intent on complicating life. . . . Merely finding the right man to talk to was a day's work for the uninitiated."[17] Bureaucrats could dismiss the protesters, but members of the city council were forced to deal with constituents, since they periodically had to run for office. Additionally, their meetings were open to the public and afforded time for constituent remarks. Weaver took advantage of this opportunity. In her presentation to the council, she particularly focused on the fact that residents had not been notified of pending construction. After council members voted unanimously to require resident notification, the bulldozers "slunk back down the hill," wrote John Weaver. In appreciation, the couple's neighbors presented Harriett with a pair of gold earrings.[18]

The apparent victory again turned out to be temporary, and Harriett soon was joined by many other mountain residents, men and women, who protested not just residential construction but a proposal to turn mountain roads into freeways and plans for shopping centers and hotels throughout the hills. There were other issues as well, including fire and flood protection. The floods of early 1952 served as a catalyst for creation of the Federation. A series of devastating fires fueled further concerns about overbuilding on brush-covered hillsides, particularly in years when Santa Ana winds swept through the canyons during periods of extremely low humidity, usually in late summer and autumn.

The most destructive occurred in November 1961, when a raging inferno swept through Bel Air, destroying nearly five hundred pricy homes, including those belonging to Burt Lancaster, Joe E. Brown, and Zsa Zsa Gabor, and the residence leased by the former vice president and future president Richard Nixon and his family. It "roared through the . . . mountains from Beverly Glen westward through Bel-Air, flew across the freeway at Sepulveda Canyon, and raced on through the Brentwood hills and toward another fire in the Malibu country," wrote Lillard, who lost his own home in the blaze.[19]

Through the 1950s it was men who mostly led the battle to control development. In the early 1960s, as in other nascent environmental campaigns, women moved into leadership positions through hillside federations and other organizations. Virtually all moved to Los Angeles from somewhere else and were married to successful men, though at least two of the women ultimately divorced. One of the divorcees, Lillian Melograno, ran a successful real estate business. Like others before and after, the women consistently faced allegations that they were elitists. Their critics seemed to have ample cause to characterize them as such, since their early public comments and writings revealed them to be mostly focused on maintaining their lifestyle and their neighborhood's exclusivity. As Harriett Weaver wrote in the early 1960s, "It never occurred to us that the [proposed developments] might not be of the same general size and personalized style that had long made all of us so satisfied with our hillside homes, or that the houses might in any way interfere with the pleasantly unhurried and reasonably spacious way of life we've adopted."[20] Another resident, Betty Dearing, specifically took aim at planned condominiums, declaring, "Townhouses are for towns—for crowded city re-developments where walking is the mode of transportation—where blighted slum districts are being refurbished ... not where scarce suburban land is being built upon for the first time."[21]

These statements may have been tone-deaf and classist, but they raised alarms about a serious issue with widespread ramifications. Over the years corporations and wealthy individuals from as far away as the East Coast had bought up thousands of acres throughout the mountains. They included the Janss Corporation, which built much of Westwood and Holmby Hills; the Lantain Company; U.S. Plywood Company; and Clinton Murchison Jr., owner of the Dallas Cowboys football team. The corporate landowners purchased the land to make a profit and cared little about conserving natural resources. "One day the chaparral on the south slope is alive with quail and squirrels," wrote John Weaver. "Next evening it has been clawed naked by bulldozers, and the cream-colored yucca stalks have been turned into black power poles."[22]

Developers had little reason to concern themselves with challenges to their unimpeded march through Southern California, particularly from "housewives." Long experience suggested that public officials would continue to support them. They were insiders who boasted that they brought needed tax revenue to cities and counties. "The homeowner is an outsider, a stranger at the feast," wrote Harriett Weaver. But women of the Santa Monica Mountains refused this appellation. Their efforts to stop development began with piecemeal campaigns to slow or stop construction but ultimately focused on removing land from prospective builders via creation of a sprawling network of parks that, together, would form the largest urban recreation area in the nation, spanning more than 150,000 acres.[23]

The women came to their environmental activism from different directions. Several were leaders of canyon associations; others were avid mountaineers and Sierra Club adherents. Some came early to the movement and spent only a few months or years. Others, including Weaver, began with one objective—saving their neighborhoods from developers—then focused on a single issue: the relationship between development and brushfires. While all of the women played important roles in the campaign to save the mountains, arguably the most significant member of this coterie was Susan Barr Nelson, who led the parks campaign from the beginning and came to be known as "Mother of the Santa Monica Mountains."

In some aspects, Nelson resembled her counterparts in numerous white, middle-class, female-led environmental campaigns of the time. Born in New York in 1927 and reared in Los Angeles, she was college-educated, with bachelor's and master's degrees in political science from UCLA. She was married and the mother of four when she began her work, though she later divorced. Unlike many of her colleagues, however, Nelson was neither deferential to authority nor concerned about how others viewed her. She spoke her mind and took no prisoners. She was driven, intense, impatient, and indefatigable. Some described her as "abrasive."[24]

"Bullshit" seems to have been one of her favorite words, since she scrawled it—in large, looping letters—across official documents, meeting agendas, and newsletters. She disdained "do-goodism" and preferred action over making lists. "Lists are not my idea of saving parkland," she said at one point. "I do not give a damn about lists." And she had little use for those who worked for public officials. "They are errand boys," she wrote. The few female staff she dubbed "errand girls." Nelson was "interested in the park, not some staff agenda to make themselves look good." At one point she told a reporter, "Sometimes I think my whole life has been wasted chasing a bunch of spectators around the mountains."[25] Nelson's brusqueness—"She could explode at you," said one colleague—may have led fellow activists to label her "difficult," but her refusal to take no for an answer or to back down in the face of frustration and defeat made even powerful politicians loath to cross her, at least publicly.[26] Eventually she wore down many of her opponents, who gave her most of what she wanted so they would stop hearing from her.

Planning for the park system began in 1963, when it became clear that protesting individual developments had little impact on turning public officials away from their strong support of growth. A small portion of the mountains lay in Ventura County, but activists focused their efforts mostly on

Los Angeles city and county officials. A year earlier the Los Angeles Planning Commission had estimated that the mountains could sustain approximately 43,300 new single-family homes and 2,200 multiple dwelling units. The new homes would net the city approximately $7 million in revenue annually. Subsequently city officials announced their intention to craft a Master Plan for the Santa Monica Mountains. Residents worried that the pending plan would provide a blueprint for ever more development.[27]

In response, they created an organization, Friends of the Santa Monica Mountains, Parks and Seashore, which aimed to craft a different narrative: that the mountains belonged to the people of Los Angeles and should be preserved for their use in perpetuity. Nelson, a founding member, soon moved to the forefront. Although she was often described as president in news stories, for many years the group's letterhead listed her as executive vice president and Ralph Stone, an environmental engineer, as president. Nelson recognized early on that to succeed, mountain activists needed to get beyond individual interests, reflected in the numerous canyon organizations. She called this approach "a save-my-canyon-but-the-hell-with-the-next-canyon attitude" and instead advocated "talking to people in every canyon and say-ing, 'Look, the way we protect our quality of life is to offer it to everybody.'" She argued, "This must be a program that involves something for the whole city." A large urban park fit the bill, particularly since the region had only one park in that category, Griffith Park. Gifted to the city in the 1890s by a wealthy miner named Griffith J. Griffith, it spanned 4,300 acres and featured an observatory, hiking and equestrian trails, a zoo, and playgrounds.[28]

The notion of preserving land for parks in the Santa Monicas was not new. In fact it dated back to the early twentieth century. Frederick Olmsted Jr., the son of the famed landscape architect and a California conservationist, unsuccessfully lobbied for parks in the 1920s, as did a wealthy mountain resident, Sylvia Morrison. In 1944 the estate of the famed cowboy comedian Will Rogers donated his 160-acre Pacific Palisades estate to California for a historic state park. And nine years later the Carrillo family trust donated 2,500 acres in Malibu for Leo Carrillo State Park on land owned by the late actor and preservationist. Born to a pioneer California family, Carrillo was probably best known for his role as Pancho, sidekick to the Cisco Kid in the

1950s television series of the same name.[29] By the 1960s, however, parks had moved far down the list of preferred uses, at least among public officials. As one area resident, Ellie Oswald, reminded preservationists, as late as 1950 parks remained on the official radar, but by the mid-1960s they had been replaced by housing.[30]

Friends of the Santa Monica Mountains sought to significantly expand the amount of land dedicated to parks, though at the beginning it was unclear exactly what form this would take. They began by recruiting members, fundraising, and widely publicizing the campaign. Early supporters included actors James Garner and Eva Marie Saint. Donations paid for ads on radio and television. In December 1963 local television station KNXT aired a segment focused on the mountains. "About 25 percent of the area of the city of Los Angeles lies in the foothills," proclaimed the narrator. "Hillside living has become a spectacular feature of the city . . . so attractive, in fact, that bulldozers continue to knock off the tops of knolls, to gouge out winding roadways, and to level off building sites for more and ever more homes overlooking the city below." But "one vast tract remained undeveloped—about 90 square miles of wild, untouched land." While public officials favored "houses, shopping centers, and industrial parks, the mark of a mature city is its parks as evidence of civic pride."[31]

The master plan, released in July 1964, provided park advocates with additional ammunition. City leaders tried to spin it as neutral with regard to development. "Rather than [providing] rigid rules on types and locations of individual buildings," its stated goal was only to "exercise a general control of [mountain] development." This would occur by requiring builders to set aside some buildable land for recreational uses. Mountain activists were not fooled. Developers got to choose which land to set aside, and the plan envisioned an eventual mountain population of 141,000.[32]

Builders were not fooled either. Plans for several large developments were soon announced. The Janss Corporation proposed 1,750 condominiums above Forest Lawn Cemetery in Glendale; the U.S. Plywood Corporation proposed 1,100 units on three hundred acres in Beverly Glen; the Spindler Realty Corporation planned 275 homes on a sixty-six-acre lot, plus a shopping center; the Lantain Company bought thousands of acres between Topanga

Canyon and the San Diego Freeway and planned a housing development spanning twenty square miles.[33]

Nelson, along with women from several canyon organizations, went on the attack. They immersed themselves in the details of every proposed development and attended every Planning Commission and City Council meeting where developments appeared on the agenda. Nelson inundated officials with letters and lobbied fiercely for an invitation to testify at public hearings. In one letter to Los Angeles Planning Director Calvin Hamilton, she wrote, "I am going to be very cross and push that invitation." Hamilton agreed to her request.[34]

Others pushed back as well. After city planners tentatively approved Spindler's Conditional Use Permit in 1965, Betty Dearing, a leader in the Federation of Hillside and Canyon Associations, wrote to every fellow member and asked each to pen letters as well. In her own single-spaced, two-page letter to members of the city council, Dearing listed objections: the developer had skirted existing zoning regulations, created conditional use permits for private gain, and planned to preserve useless land as open space, while using valuable land for apartments and condominiums. Council members ultimately turned down the proposed development.[35]

After Allied Canyon Company proposed five hundred homes and condominiums on two hundred acres in Benedict Canyon, Association leader Lillian Melograno asked the Planning Commission "to pass a resolution denouncing the granting of any variances." The Commission initially postponed its ruling and gave Melograno a month to prepare a report on the project. "It was the first time property owners hired their own experts," she recalled later. "We had a geologist, an engineer, and a traffic consultant, and we did a house-by-house survey of the area to show how many children there were, how many cars, what schools the children went to. You buy a whole neighborhood when you buy a house." The project eventually was shelved. The women also fought U.S. Plywood's effort to build nearly a thousand condominiums in the Beverly Glen neighborhood. In response, the company filed a $2 million lawsuit against homeowners groups, charging "malicious interference." The lawsuit was eventually dismissed. In an interview, Melograno described her early days as an activist: "I was so innocent!

It was so frustrating. I used to get furious. Finally, I decided that I either had to cut down on the emotions or get ulcers."[36]

In 1966 women from an array of organizations pressed for a Los Angeles County Grand Jury investigation into the cozy relationship between public officials and developers. The resulting report essentially stated the obvious: "Influence can, has been and in all probability will be exerted through the medium of campaign contributions, political obligations and friendships." At one point Nelson asked, "Is it any wonder that a majority of citizens are concerned about the character of a city government that supposedly represents them?" Betty Decter, a founder of the Beverly Drive Property Owners Association, said, "Once you become aware of what's going on in this city and what it's costing you as a homeowner and as a taxpayer, you can't turn your back."[37]

The women also promoted the candidacies of politicians they viewed as sympathetic to preservation. They phoned, wrote letters, knocked on doors, and attended meetings. In 1965 their efforts resulted in the election of Marvin Braude, a Brentwood resident and fierce environmental advocate, to the Los Angeles City Council. It was the first of what would become eight terms for Braude. Three years later Barry Goldwater Jr., a mountain resident and son of the Arizona senator and 1964 Republican presidential candidate, won the first of seven congressional terms. The women sought assistance from legal experts; for instance, Nelson recruited Robert Jesperson, executive director of the Conservation Law Society, to identify lawyers willing to file actions against builders over zoning and other violations "which [the women felt] would materially damage a proposed two thousand acre state park."[38]

The activists worked to ensure that funds from a $150 million state parks bond measure passed by California voters in 1964 went to the Santa Monica Mountains. In fact 30 percent of the money was earmarked for mountain parks, resulting in the establishment of the fourteen-thousand-acre Point Mugu State Park in Malibu. Though she cheered the added acreage, Nelson bemoaned the "often lonely and unsupported" campaign to remove land from developers. She denounced the lack of action, even as she enumerated her contributions to the ongoing campaign. These included obtaining TV coverage of events and making appearances at hearings: "Last March I went to Sacramento to meet with [the director of the California Parks Service] William Mott. Thursday, I attended the state Highway Commission in Sac ... and I did testify."[39]

By the end of the decade, pro-development members of the Los Angeles City Council and Planning Commission had come to recognize the mountain advocates as formidable opponents. As a result they abandoned their long-held desire to add 100,000 new residents to the mountains and instead proposed preserving "unique natural qualities and features of the mountain area." By 1970 they required developers to submit traffic analyses prepared by engineers before starting construction.[40]

The activists made inroads elsewhere as well. Harriett Weaver's work on preventing brushfires resulted in 1968 in Los Angeles county and city ordinances mandating brush removal within one hundred yards of any structures. Failure to comply brought stiff fines and the prospect of steep hikes in insurance premiums. Weaver targeted developers in her public remarks. "Subdividers claim proudly that the development of a formerly wild area makes it safer because it reduces the amount of brush," she wrote in a newspaper opinion piece, "but experts point out that the removal of one hazard must be weighed against new ones ... and they are legion." She cited development-linked problems: landfill that caused a buildup of water that led to flooding; power lines that fell in high winds, setting off sparks that fueled fires; construction equipment that tore up brush and sent debris flying; fire from backyard barbecues, motorcycles, and cigarettes. "Fortunes are to be made [in the hills], and powerful interests want to make them— builders and subdividers, money lenders and lobbyists, even public officials," she wrote. Hillside residents, on the other hand, sought "a place to live in peace on an unspoiled bit of land they call their own." It took another two decades to implement another of Weaver's goals: banning highly flammable wood shake roofs from mountain homes.[41]

It was a three-year battle over Mulholland Drive that pushed the parks campaign into high gear. State lawmakers had proposed legislation to turn the iconic two-lane road into a freeway, with a sixty-mile-per-hour speed limit. If the Mulholland effort succeeded, it would give developers access to property throughout the mountains—even in the most remote areas. Large landowners, including developers, owned more than a third of the property available for widening. A freeway also would also mean turnouts, gas stations, and convenience stores for drivers. And it threatened to displace wildlife.

Residents strenuously objected to the plan. Mulholland, they declared, was "utterly unique in Los Angeles," providing a panoramic view of the city and surrounding area. Melograno argued that the planned freeway would "destroy the entire character of the Santa Monica Mountains due to the tremendous grading involved." The mountains, she added, would "be leveled to a plateau to accommodate such a road." In a letter to the *Los Angeles Times* she wrote, "This is our one and only chance to preserve a unique and magnificent scenic wonder. The future of a beautiful Los Angeles is at stake."[42]

Ironically property owners themselves seem to have borne some responsibility for the freeway proposal. In the early 1960s they had proposed a special assessment district to improve—though not widen—Mulholland Drive. The Los Angeles City Council denied the request and instead petitioned the state to have Mulholland named part of the state's scenic highway system. State lawmakers approved the designation, but also proposed widening the highway to four lanes and doubling the speed limit.

Nelson made dozens of trips to Sacramento to lobby against the proposed legislation. She touted her dedication, having spent three years testifying against the proposed freeway before the state legislature. In her testimony to state lawmakers, she declared that Mulholland's expansion would mean the nearly total leveling of the mountains between the San Diego Freeway and Topanga Canyon. She warned one legislator, "You are going to be so tired of hearing from me before we are through. I am leaving come June for a camping trip with my family, so after June 29, you will not have to hear my voice. However, until that time, let me give you some of my thoughts."[43] Eventually lawmakers let the bill die.

The fight over Mulholland Drive brought new activists into the battle, including two who would come to work closely with Nelson. Margot Feuer lived in the hills above Malibu with her husband and three children. A graduate of Wellesley College, she was an early leader in Stamp Out Smog and active in the Sierra Club. She later explained why she joined Friends of the Santa Monica Mountains, Parks and Seashore: "I looked around at what I was in the middle of and I figured, gosh, the idea of a park is a beautiful idea." Her Sierra Club connections made her an effective and respected lobbyist.[44]

Jill Swift was a graduate of Stanford University, an avid hiker and Sierra Club member, who had trekked all over the world. In the 1960s she was a housewife and a Girl Scout leader living in Tarzana. One Sunday she and her daughter decided to hike in a canyon near Mulholland Drive. "[Even limited to two lanes,] we couldn't hear the birds, we couldn't hear each other talk," she said. "There were motorcycles and people with trucks all over. All I could think of was, 'we've got to protect this road.'" She complained so vociferously and so continuously that "[her] kids dared [her] to stop complaining and do something about it." Swift decided to organize a hike to "get people out to see what was in the mountains" and bought a new pair of hiking boots. "I put the boots on and walked around the neighborhood after dark.... In those days you didn't see too many housewives trucking around in hiking boots." She expected three hundred hikers for her first outing. Five thousand showed up. She began leading more than a dozen hikes a month, creating a grassroots movement in the process. One frequent participant told her, "We'll lead 10,000 people into the mountains and maybe 1,000 will work to preserve them." She distributed guidebooks describing the native plants and wildlife they would see.[45]

By the early 1970s Friends of the Santa Monica Mountains, Parks and Seashore was thriving. Two thousand people read the organization's newsletter. Many high-ranking officials had joined its advisory board, including California's lieutenant governor Glenn Anderson and Senators Alan Cranston and John Tunney. With the organization's help, in 1971 Joel Wachs, a strong environmentalist, defeated James B. Potter to win a seat on the Los Angeles City Council. Potter had once complimented Betty Dearing for her "ladylike demeanor" when she testified before the council. In 1973 Los Angeles voters elected Tom Bradley as the city's first African American mayor. Bradley, who defeated conservative incumbent Sam Yorty, was a strong supporter of Friends. "During the Yorty regime," Betty Decter recalled, "the Planning Commission was completely developer-oriented. We're in a much better ... position than we used to be."[46]

"If you look at just about any major park," said Russell Cahill, California State Parks director in the 1970s, "you'll find—primarily—women who did the groundwork and stayed with it, kept the pressure on, and who worked the issues to the point where they succeeded."[47] The women of the Santa Monica Mountains may have been inexperienced in guiding such a large-scale endeavor, but they had some idea what they were up against. In an interview Nelson recalled, "I remember early on saying to someone: 'You know, this is going to take a long time.'"[48]

The timing seemed fortuitous. By the 1970s a consensus in favor of environmentally conscious policies had emerged across the country. A 1963 report by the Recreational Advisory Council of the U.S. Department of the Interior had predicted the growing desire for more public spaces, taking note of "the steeply mounting outdoor recreational demands of the American people." The report recommended that all levels of government—federal, state, and local—as well as private interests become involved in acquiring park land. Most of the parks should be sited in urban areas, it noted, since, within a few decades, as many as 70 percent of Americans could live in or near cities.[49]

The report created a road map of sorts for Santa Monica Mountains activists. Congress was already in the process of creating an urban recreation area in California: the eighty-seven-thousand-acre Golden Gate Recreation Area in the San Francisco Bay Area. It included Alcatraz Island, the Presidio, and the Marin Headlands. Why not create a national park in the Santa Monica Mountains to permanently remove land from development? "If the county Board of Supervisors in the previous 20, 30, 40, 50 years had done what it should have, perhaps it wouldn't [be] necessary for the federal government to step in," said Congressman Anthony Beilenson, who represented part of the Los Angeles area.[50]

But how to sell a massive mountain park to lawmakers representing far less affluent areas of the Los Angeles basin, and who sought funds for parks in their own districts? Many viewed the mountain campaign as yet another example of the politics of privilege; some seemed poised to derail it. In response, park proponents crafted a new, more inclusive message: parks benefited everyone. For example, they provided poor families with opportunities to experience open spaces, free of charge. As one member of

Save the Santa Monica Mountains told the Greater Los Angeles Press Club in a speech, "There are too many people who can't go to expensive camps. Children play in alleys and on asphalt parking lots who we put on buses and sent at city expense to the Sierra Nevadas, neglecting all the while this last resource we have so close at hand."[51]

Another approach proffered parks as providing opportunities for city residents to escape the basin's smog. Nelson framed it this way: "One major factor [in the persistence of smog] has been the replacement of open space with concrete. Natural open space, in addition to providing recreation, is a natural smog barrier. The Santa Monica Mountains represent one of the last remaining land areas in the region. That they have been targeted for misuse and destruction is unthinkable."[52]

Margot Feuer was an expert on pollution. "If the open spaces of the mountains are replaced with housing and commercial developments," she declared, "the cleansing action of vegetation will be reduced. . . . In its place will be the emissions of pollution." She called the Santa Monica Mountains "an air shed for an already critically polluted basin."[53] Yet another strategy focused on young people. "The youth of our future generations are not going to live in a community of spacious back yards and open space," declared Marvin Braude's wife, Marjorie, a mountain activist and psychiatrist. "They are going to live in a dense suburban community."[54] Meanwhile Jill Swift continued leading weekly hikes through the mountains, garnering support from the growing number of participants.

Nelson's leadership and relationships with political figures put her in an advantageous position to guide the park campaign through the political process. "From the very beginning, it was Sue Nelson . . . whom I most trusted," said Congressman Beilenson. She was "responsible for much of the grassroots lobbying for parkland." Dale Crane, who worked for state senator and later congressman Phillip Burton, said Nelson "was constantly there and constantly working." Crane acknowledged Nelson's prickly nature: "I do remember many long arguments with Susan about what we could and could not do."[55] Another official described her as "the principal person who gave [officials] the most grief."[56] Apparently she was not alone in this regard. Russell Cahill of the National Park Service recalled that Nelson, Feuer, and

Swift "were very good to work with." However, "they didn't like being crossed. None of them liked any disagreement."[57]

Some fellow board members of Friends seem to have grown exhausted with Nelson by the early 1970s. At least two resigned their positions and blamed Nelson for fostering a hostile environment. She denied intentionally alienating anyone. "I really do not accept the reason given for yours and Betsy's resignation," she wrote to an unnamed recipient. In a fit of pique, Nelson offered to resign from the board as well, giving as her reason "the neurotic shit that I have been taking." Fellow board members seem to have accepted her resignation, which she soon rescinded. "What bullshit," she wrote.[58]

The year 1974 was the turning point for the Santa Monica Mountains effort. In January, Los Angeles planners and representatives from the city Parks and Recreation Department met to discuss logistics and potential park locations. Nelson was the only nonofficial invited to attend. That spring and summer the U.S. Senate held hearings in Los Angeles and in Washington DC on what would be called the Santa Monica Mountains National Recreation Area (SMMNRA). Senator John Tunney, a Democrat, sponsored SB1270, which called for acquisition "through direct appropriations, donations, gifts, leasebacks and other arrangements" of more than a dozen pieces of land in Zuma Beach, Malibu Canyon, Century Ranch, Trancas, and Cold Canyon.

Nelson, Feuer, and Swift all testified at the Los Angeles hearing in June. "Every time a small community plan comes up, the people will be given instructions from the local government that nothing will be in it about parks and open spaces," Nelson told lawmakers. "We go to every single community meeting, and I will tell you it takes five meetings a week, in the evening, all evening to get to people ... to sort of get them in line in terms of the total program." Feuer focused on smog. "Local and State governments have made fragmented purchases in the mountains and beaches for park area," she said. "But it will take the financial resources of the Federal government to unify the major part of the vast and significant space as an air shed for an already critically polluted air basin." In brief remarks Swift emphasized the growing number of people joining her for mountain hikes: "The number of people hiking is evidence of hunger for public parks."[59] The bill failed to advance out of committee, however, in part because of Washington's fixation on Watergate.[60]

Nelson, irate, called and wrote lawmakers in protest. A legislative assistant for Congressman Alphonzo Bell of California wrote her an apologetic note, assuring her, "Mr. Bell has been very actively pursuing this matter." That September she received another apology, this one from Sacramento-area Democratic congressman Harold T. Johnson, who admitted the bill would be postponed with no time table for new hearings. In an undated newspaper article, Nelson lamented the tortoise-like progress: "[The SMMNRA] has always been a 'people's park,' trails and scenic corridors running from Griffith Park to Pt. Mugu in Ventura County." In Washington DC, however, "the Santa Monica Mountains are often viewed as a local issue."[61] Her anger shone through another missive to members of Friends: "Our Southern California Santa Monica Mountains and Seashore project is one of the most remarkable kind of greenbelt in America and all we are getting is lip service and the run around. Do something about it. Write to the President of the United States."[62]

The national park bill may have been shelved, at least temporarily, but California voters in 1974 enthusiastically backed more state parks, passing a $250 million bond measure for state beaches, parks, and recreational and historical facilities. More than half of the money was slated for parks, and nearly $50 million for the southwestern coastal mountains—including the Santa Monicas—and the Sierra foothills. But it was not enough for Nelson, who scrawled notes across official documents, such as "Why is this in here?" and "This is not a beach!"[63] Ultimately two more state parks were opened that year in the mountains: the 12,666-acre Topanga State Park, between Malibu and Pacific Palisades, and the 4,000-acre Malibu Creek State Park, which included ranch land donated by Bob Hope and Ronald Reagan.[64]

In 1976 Congressman Beilenson brought together a group of park activists to craft a new bill creating a national recreation area. The group included Feuer, Swift, and Nelson. Congressmen Barry Goldwater Jr. and Robert Lagomarsino—a Republican and a Democrat—introduced their own legislation. Neither bill gained traction. The estimated cost—over $100 million—doused hope for quick resolution. Nelson complained, "Like coal, the Santa Monicas are there to be mined or developed—San Fernando Valley–ized. We've had a difficult time getting our elected officials involved." She blamed campaign contributions from developers.[65]

By fall 1977, however, the Santa Monica Mountains had "become one of the hottest federal park proposals in the nation," proclaimed the *Los Angeles Times*. The reporter, Robert A. Jones, attributed its enhanced fortune to intensive lobbying, widespread publicity, and new leadership on several crucial congressional committees, primarily the selection of the San Francisco Democrat Phillip Burton to head the House Subcommittee on Parks and Insular Affairs.[66] These developments—particularly Burton's involvement—led some observers to declare the Santa Monica Mountains "saved," a pronouncement that sent Nelson and her colleague Mary Ellen Strote to the typewriter. "A nice thought," they wrote in an opinion piece for the *Los Angeles Times*, "but unfortunately it isn't true." The region was still threatened by "speculators with subdivision maps and blueprints rolled under their arms." They added, "The reasons for such a long delay in winning [park] approval remain a mystery.... The opportunity will not last a decade or so this time.... We are looking at a classic case of now or never."[67]

Burton's selection proved to be a game-changer. He was an unlikely savior, since he did not represent the Santa Monica Mountains and was not particularly keen on the outdoors. "He lived a lot on vodka and cigarettes," said one official, who called Burton "a hard-living populist and social activist." Burton knew the Santa Monicas, however, because he and his wife, Sala, had honeymooned there decades earlier.[68]

Considered one of Congress's most brilliant and canny legislators, he was single-minded and ruthless when it came to getting what he wanted. And what he wanted in late 1977 was to pass a parks bill, partly because he had narrowly lost a battle to become the House majority leader and wanted to demonstrate his ability to shape a political agenda. And also, according to his biographer John Jacobs, to "show that he was smarter than anyone else."[69] Burton knew, however, that few members of Congress from other states would back a measure giving California millions of dollars. So his bill—157 pages long and asking for $1.8 billion—included parks in virtually every state. He also sought to defuse problems with inner-city lawmakers by including $725 million to upgrade playground equipment in their districts. His strategy was "that if he had something for everybody, everybody would vote for it," said Arthur Eck, the legislative affairs specialist for the National Park Service at the time. "And in the end, his process was proved right."[70]

Burton's bill allotted $155 million for a national recreation area in the Santa Monicas: $125 million to create a sprawling national park, $500,000 for park development, and $30 million in federal grants to California for use to purchase land in the mountains. Additionally it preserved Mineral King, a pristine wilderness in the Sequoias, something for which the Sierra Club had long advocated in the face of efforts by the Walt Disney Corporation to turn the area into an expensive ski resort. The bill easily passed the House but ran into trouble in the Senate. So Burton incorporated items in pending Senate bills into his omnibus measure. In the end, ninety senators had items in his bill, which passed in October 1978.[71] It was the largest parks legislation ever enacted by Congress and "a watershed moment in environmental and natural resource policy-making," said Dale Crane, an aide to Burton.[72]

SMMNRA would be administered by the National Park Service, though it would not be called a national park. That designation generally went to isolated areas, such as Yosemite, Yellowstone, and Glacier national parks, not urban areas. State parks in the mountains would be included within the boundaries of the national park. After President Jimmy Carter signed the legislation in November 1978, more than one hundred park activists gathered in Nelson's back yard for "a potluck victory celebration." A *Los Angeles Times* reporter covered the event. "The list of victories by the conservationists fails to hint at the work that went into achieving them," the reporter noted: "drafting of legislation, then persuading legislators to introduce it; hearings and committee meetings; formation of study commissions, lobbying in Los Angeles, Sacramento and, eventually, Washington." The women could finally find humor in their many challenges, which had included educating dense and oblivious lawmakers about the mountains. One legislator had asked, "Where the heck are the Santa Monica Mountains?" Another, when accused of being "crooked," replied, "You're dead right—but you don't have to say it in public."[73]

But passage of the bill was only the first step. Debate ensued over where new park land should be located. Land was more expensive in the eastern, more developed part of the mountains. But remote land, though cheaper, generally was not close to existing state parks. And some open space available for parks abutted existing residential neighborhoods. Some landowners proved willing to sell or lease their land, while others refused to do so. "We knew from the beginning it would be a patchwork," Congressman Beilenson said.[74]

Longtime park advocates also proved problematic. Some got to the point "where they didn't want the grass walked on," said Linda Friedman, an aide to Beilenson. "Others were for bringing in busloads of kids from Watts." She too usually deferred to the expertise of Nelson, as well as Swift and Feuer, who hosted some of the planning meetings in their homes. All three "wanted wilderness left the way it was." Meanwhile Mayor Tom Bradley of Los Angeles was "very interested in seeing places that could be destinations with people who didn't have cars." So he proposed bus stops at park entrances.[75]

Jurisdictional disputes also emerged. The federal government would oversee the recreational area, but the state Parks Department and city and county officials also had prominent roles to play. Corporations still had not given up. One development company had plans for a subdivision next to Point Mugu in Ventura County. They were "too far along" to stop it, company representatives told public officials. Developers were "still pretty aggressive," said Parks Director Cahill, but they knew it was a losing battle by this point. Then there were property rights groups. They acted as shills for corporations but tried to hide their intentions, claimed Joe Edmiston, director of the Santa Monica Mountains Conservancy. By the late 1970s the California Coastal Commission had been added to the mix, and it too weighed in on how coastal lands were to be used.[76]

Park boundaries became a sticking point as well. Friends of the Santa Monica Mountains drew a map encompassing all of the land they thought should be included in the national recreation area. "I think they used a very wide felt-tip pen," said one participant. "At one point, the Santa Monica City Hall was included." After many heated discussions, Congressman Burton brought all of the participants together. "Burton says, 'Now, what's the boundary?' The National Parks Service had a boundary and Sue Nelson . . . she had a boundary, and our boundary was larger in some areas than others. We started arguing, 'I want to include this, I want to include that.'" Burton grabbed the pen and drew the boundary, which included most of the land requested by the others.[77]

In 1977, as odds grew in favor of SMMNRA, the California legislature created the Santa Monica Mountains Comprehensive Planning Commission, an advisory body. Chaired by Los Angeles City Council member Marvin Braude, the fourteen-member group was tasked with creating (yet another) mountain master plan, this one "consistent with preservation." Nelson did not serve on the commission, but she offered constant input via numerous phone calls, letters, and personal visits. The commission thus approached the process with the understanding that the recreation area "was essentially a Sue Nelson thing," said one member.[78]

By summer 1979 the Santa Monica Mountains Comprehensive Planning Commission master plan was complete. The report described the mountains as "the only unspoiled coastal range in the midst of a major metropolitan area with both coastal and mountain ecosystems" and noted that "a history of fragmented political jurisdictions [had] led to piecemeal development." In the end, Nelson and her colleagues won virtually every battle. Much of the mountain terrain was deemed too steep to build on, thus building should be limited, the commission stated. With a nod to Harriett Weaver, commissioners took note of the potential for deadly fires, as well as floods and earthquakes. The land was unstable and soil erosion a significant factor. Not surprisingly, commissioners had zeroed in on air pollution. Without the mountains as barrier, smog would be even worse in the Los Angeles basin, they declared. And more mountain development would cause more smog everywhere. The master plan also highlighted the Native American history of the region, recommending that the recreation area include a Chumash history center. And it called for no new "cross-mountain" roads to be built and advocated for buses to carry park-goers to their destinations to minimize the impact of cars.[79]

The following year the state legislature created the Santa Monica Mountains Conservancy to "buy back, preserve, protect, restore and enhance treasured pieces of Southern California."[80] The first state land acquisitions of the new era began with Rancho Sierra Vista and the Paramount Movie Ranch. The 850-acre Rancho Sierra Vista sat adjacent to Point Mugu State

Park and was donated by Richard Danielson, who earlier had donated 5,585 acres for Point Mugu. The National Park Service purchased the 2,700-acre Paramount Ranch, the location for many TV and movie westerns, including *Gunsmoke*, *The Rifleman*, and *Gunfight at the O.K. Corral*. But the federal acquisition effort still faced headwinds. The 1980 presidential election of Reagan brought James G. Watt as interior secretary. Watt favored opening public lands to oil drilling and development, and he favored offshore oil drilling in California, including in Santa Monica Bay. He declared the Santa Monica Mountains "too fragile" to accommodate a large recreation area. During Watt's tenure, the federal government significantly scaled back the budget for national parks.[81]

Nonetheless progress on the overall recreation area proceeded apace. In 1982 the National Park Service completed a "general management" plan for the area. It featured dozens of "activity sites" for visitors' use, including hiking, camping, picnicking, horseback riding, and whale watching, as well as sites for cultural and educational facilities. Over the next nearly three decades, a patchwork of parkland emerged, spanning the coastal lands on the western edge of the mountains to the freeways skirting the mountains in the east. Among the twenty-seven parks within the recreation area were several—including the Paramount—with connections to Southern California's cultural community. In 1993 Barbra Streisand donated her twenty-four-acre Malibu estate to the Santa Monica Mountains Conservancy. A decade later the National Park Service purchased Bob Hope's ranch just east of the city of Thousand Oaks. It was "the crown jewel," said Conservancy Director Edmiston. That same year the Conservancy purchased the nearly three-thousand-acre Ahmanson Ranch in Calabasas. Howard Ahmanson was a major benefactor to Los Angeles theater and the arts.

Today the national recreation area sprawls across 155,000 acres, encompassing five zip codes. The state Parks Department owns 42,000 acres, the National Park Service nearly 24,000. Local agencies, private property owners, and universities control much of the rest. More than a half-million people visit the area each year. In 1999, Arthur Eck, then superintendent of the SMMNRA, said in an interview, "When all is said and done, you will have spent half as much [on the Santa Monica Mountains] as was spent for Redwood

National Park . . . [for a park] that serves anywhere from 100 to 1,000 times as many people on a daily basis."[82]

Saving the Santa Monica Mountains from corporations and developers did not mean the end of activism for most of the women who stood on the front line of the battles. Harriett Weaver spent most of the 1970s serving on the Los Angeles Countywide Citizens Planning Council, and afterward she continued working with firefighters and city officials to ensure compliance with the brush fire ordinance. "Clear the brush . . . because if you don't we will and you'll pay the bill," she wrote. When Weaver died of melanoma in 1988, Jerry Fields, past president of the Board of Fire Commissioners, told those gathered for a memorial service, "This special woman had an inordinate amount of courage, fortitude and just plain guts. There is an old adage that goes, 'you can't fight city hall.' Harriett Weaver did fight city hall and did win. In doing it, she showed us a better way to protect the hillsides of our city."[83]

Jill Swift died in 2008 of multiple myeloma. She was seventy-nine. As her obituary writer noted, she built grassroots efforts "one hike at a time." She "brought people in direct contact with the area's beaches, trails, flora and fauna. She led thousands of people in the mountains."[84] Margot Feuer died in 2012 at the age of eighty-nine. In an interview that year she said, "The public is so often chastised for not really hanging in. [The decades of work to create SMMNRA] reaffirms the need for hanging in. All those people who worked—for that length of time—did it."[85]

Susan Nelson continued working on mountain projects for the remainder of her life. She continually pressed for the acquisition of more park lands, and she led projects to map the wildlife and plant life and worked to uncover Native American archaeological sites. In the 1980s she moved to Echo Park, near downtown Los Angeles, and continued her activism there, adding environmental justice to her repertoire. For example, she fought against a proposed freeway in the East Los Angeles city of El Sereno, arguing that it would displace many low-income residents.

For years Nelson had prodded Leonard Pitt, a history professor at California State University, Northridge, to write an oral history of the campaign to save the Santa Monica Mountains, and she donated funds for the effort. Pitt interviewed more than a dozen people involved in the effort. He dedicated the finished work to Nelson, whom he dubbed the "Mother of the Santa Monica Mountains."[86] Nelson did not live to see the finished project. In May 2003 she was crossing Sunset Boulevard on foot when she was struck by a van and killed; she was seventy-six. Her former colleagues in the mountain preservation effort recalled her unstinting activism. "She was just everywhere," said Sarah Dixon, a member of the parks advisory commission. "She was an example of someone who worked, never for personal gain, but just because [something] needed to be done." Congressman Beilenson said, "She was very much the single greatest private force behind our working to create the park."[87]

In November 2018 Santa Ana winds again blew with gale force through Southern California. Fire started on the flatlands and soon swept through the canyons of the Santa Monica Mountains. Dubbed the Woolsey Fire, before it was contained it had burned nearly 100,000 acres, including more than 80 percent of the federal recreation area. A year later most of the area had been reopened. "If the ridges and canyons that now make up the recreation area were packed with the high-density development once envisioned by planners," wrote a *Los Angeles Times* reporter, "the outcome of the Woolsey fire, in human lives, could have been exponentially worse. Conservation sometimes conserves more than we realize."[88]

5

FIG. 5. Marie Harrison, a San Francisco environmental justice advocate for Greenaction and other organizations, protests in front of a PG&E facility in 2004. Source: Mike Kepka/*San Francisco Chronicle*/Polaris.

ENVIRONMENTAL JUSTICE
The Politics of Survival

Pamela Tau Lee became an activist in the late 1960s as a student at California State University, Hayward, south of Oakland in the East Bay. Lee was a San Francisco native. Born in 1948, she spent her first seven years in Chinatown. When her family moved to an integrated neighborhood, she became one of very few non-Anglo students in her school. Her parents told her that, being Chinese, she needed "to work harder." Despite this admonition, she was not a particularly good student, Lee told an interviewer much later.[1] She attended San Francisco Community College before transferring to Hayward to major in sociology.

It was a heady time to be young and politically involved, particularly in the Bay Area. Anti–Vietnam War protesters burned draft cards, and Native Americans occupied Alcatraz. The nascent women's movement was tackling gender inequities. At Hayward, as at other colleges, students demanded relevance in their studies. Lee's "political awakening" occurred when she joined a multiethnic group of students at Hayward. After graduation she enrolled in a program called Teacher Corps, but ultimately dropped out because she realized she "wanted to be in the streets."[2]

In the late 1970s she moved back to San Francisco and became a labor organizer, helping to unionize hotel employees, most of whom were minorities. Since she wanted to gain a better understanding of their working conditions, she took a job as a hotel room cleaner at the San Francisco Hilton. "I cried everyday," she recalled. "It was a horrible, horrible place to work." Managers hit, shoved, and insulted workers. The industry was rife with abuses. Maids had to clean eighteen rooms a day and pay for their own food. To recruit union members, Lee went door to door in the Tenderloin, a run-down section of San Francisco, where many of the workers lived. The buildings were dirty and dilapidated, doorbells were broken, and the hallways were soaked with urine.[3]

In one campaign to unionize workers at the Mark Hopkins Hotel, balloting took place at the Top of the Mark, the iconic cocktail bar on the top floor of the elegant hotel named for one of the "Big Four," men who built the western portion of America's first transcontinental railroad. The significance of the location was not lost on Lee or the other organizers. If the opulent and forbidding ambiance was designed to intimidate workers, however, the tactic failed. In the end, they voted overwhelmingly in favor of the union. Lee recalled, "[I] got so dizzy, I almost passed out right then and there."[4]

In the 1980s Lee turned her talents and energy to an emerging movement called environmental justice. It reflected the growing recognition that "struggles for environmental justice and struggles for human dignity" were inexorably linked.[5] The movement was partially modeled on Martin Luther King Jr.'s 1968 Poor People's March. Mainstream environmentalism had made many gains since the 1960s, but it had hardly touched the lives of the people Lee met in industrial areas such as Richmond and other working-class towns in the shadow of refineries, chemical plants, incinerators, and hazardous waste facilities. Many residents were poor people of color; some were immigrants who spoke little or no English. As such, they had extraordinarily limited access to those who made the decisions that impacted their lives. They could not attend public meetings and anticipate the level of respect or deference experienced by the women of Save the Bay or Stamp Out Smog. And they could not telephone public officials and expect quick—or any—responses.[6]

But political activists such as Lee and others in working-class and minority communities had watched the mainstream movement emerge and take root, and they witnessed new strategies and tactics unfold over time: from middle-class housewives attending meetings, lobbying lawmakers, and raising money to more direct community action. The antinuclear movement played a role in this process, for it opened the door to "the issue of toxics" and "resulted in the emergence of thousands of local groups focusing on hazardous wastes."[7]

The Clamshell Alliance on the East Coast and Abalone Alliance in California may have provided a framework. Clamshell was formed in 1976 to protest the Seabrook nuclear power plant, to be sited on marshes containing clam beds in New Hampshire. It included people from all walks of life—teachers, writers, factory workers, community organizers—who practiced nonviolent civil disobedience via town hall meetings, picketing, workshops, and site occupations. These actions often ended in arrests. It was a diverse group, with Native Americans, women's rights activists, young, middle-aged, and older people.[8]

The Abalone Alliance was formed a year later, to protest the construction of the nuclear power plant in Diablo Canyon. The group linked nuclear power to an authoritarian military-industrial complex. Named for the abalone beds that would be destroyed by waste from the plant's reactors, it was much less diverse than Clamshell. Most members were white and middle class, but women did play prominent roles. Many came from the group Mothers for Peace, based in San Luis Obispo, that got its start protesting the Vietnam War. Abalone's first major action took place in August 1977, when Diablo was in the early construction phase. Forty-six protesters were arrested for trespassing on PG&E property. They placed abalone shells on the road leading to the construction site. The following year three thousand people showed up to picket PG&E in San Francisco, and twenty-five thousand attended a "Stop Diablo" protest. That same year, a massive rally at Camp San Luis, a home of the California National Guard, drew forty thousand protesters, including Hollywood celebrities and Governor Jerry Brown, who touted himself as the state's—and the country's—first environmental chief executive. In 1981, despite earthquake fears, the Nuclear Regulatory Commission approved low-power testing at Diablo, sparking two weeks of continual protest in September of that year.[9]

Judith Evered traveled to the Diablo site from her home in Santa Barbara with a small contingent of women. The cancer death of her seven-year-old son propelled her toward antinuclear activism. The Evered family had lived near a nuclear plant in England, and Judith believed it had caused his fatal illness. At the construction site, thousands of protesters gathered in separate groups. One set up a staging area next to the fence that separated Diablo construction from the nearby community. A second group circumvented the staging area and approached the site from the surrounding countryside. A third arrived by sea in small boats. When the word came from group leaders, Evered and her contingent scaled the fence. They were quickly arrested and taken to a gymnasium at the California Men's Colony, a nearby maximum security prison. The backcountry group managed to temporarily evade police; the contingent in boats made its way to the rocky shore, swimming the last portion.[10]

Over the next few days nearly two thousand protesters were arrested, along with members of the media covering the blockade. Evered and her fellow arrestees spent six days waiting to be taken to court for arraignment. Police had been polite at the beginning, she noted, but over time they became impatient and angry with the protesters, putting them in chokeholds, twisting their arms behind their backs, hitting them, and tossing them into police vehicles. After a week or so, most protesters had been arraigned and released on bail. Many went home. Others, including Evered, went back to the protest site. When the blockade eventually ended, a few organized a sit-in at PG&E's San Luis Obispo headquarters.

The use of pseudonyms by many protesters slowed arraignments. Several used the same names, such as Sojourner Truth, a former slave, suffragist, and early feminist, and Karen Silkwood, an Oklahoma activist who had raised alarms about the mishandling of plutonium at the chemical plant where she worked; en route to a meeting in 1974 with a *New York Times* reporter, her car ran off the road and Silkwood was killed.[11] Some Abalone Alliance members professed embarrassment about using the name Sojourner Truth, since the group was so monochromatic.[12] One acknowledged the disconnect and said, "I think that there are a lot of people in this world who have issues that deal with their own survival." She pointed out that "white people created this nuclear mess," while the poor and people of color "suffered disproportionately from chemical and/or radiation pollution." Members themselves discussed their lack of diversity and decided to "proactively go into communities and invite participation in the anti-nuclear movement."[13] Though it did not stop the construction of Diablo, Abalone and other groups played a significant role in turning members of the public against nuclear power plants in general.

By the 1980s the time seemed right for a new kind of environmental movement, based on the "relationship of [the] environment to racism and occupational health and safety." Over the previous two decades, mainstream environmentalists had forced public officials to grapple with seemingly intractable issues, including mountains of hazardous waste products—plastics, solvents, chemicals "dumped into pits, landfills, or waste ponds, or discharged into surface waters, deep wells, or the ocean."[14] But the toxic materials still needed to be deposited somewhere, so industry and government entities had turned to poor and minority communities.

They were not embarrassed to acknowledge this decision, or their motives for making it. In 1984 the state of California hired a public relations firm to investigate how residents in various neighborhoods might respond to having waste incinerators built near their homes. Its unsurprising conclusion: "Middle and upper class socioeconomic strata possess better resources to effectuate their opposition. . . . Conversely, older people, people with a high school education or less are least likely to oppose a facility."[15] And a study published in 1987 revealed that "race was the most significant variable associated with the location of the waste facilities, with large numbers of African Americans and Hispanics, often in overrepresented numbers, living in communities where such sites were located."[16]

The disparate treatment was reflected in the politics of pesticides. Rachel Carson's seminal book *Silent Spring* had fueled a nationwide move to ban DDT and other dangerous chemicals, yet farmworkers continued to labor in fields saturated with them. They drank tainted water and carried the residue home on their skin and clothing. Constant exposure led to birth defects, chronic illnesses, and deaths from cancer. No amount of lobbying by farmworker advocates moved the needle. In fact quite the opposite. In September 1988 United Farm Workers leader, Dolores Huerta—a strong advocate of banning pesticides in the fields—along with hundreds of allies marched in front of a San Francisco hotel where Vice President George H. W. Bush, the Republican presidential candidate, headlined a $1,000-per-plate dinner. Police ordered the marchers to keep walking. Huerta apparently did not move quickly enough, and police began beating her with batons. She was so badly injured that she was hospitalized and had to have her spleen removed. The incident proved embarrassing for the city, partly because of Huerta's stature and because some city officials, including Mayor Art Agnos, supported her cause.[17]

Anger over blatant inequities fueled action among residents of impacted communities. Residents in Houston, Texas, alleging civil rights violations, sued a solid waste management company over its decision to place a facility in their neighborhood. In North Carolina, protests erupted over toxic waste dumped along a highway. And in a mostly African American public housing project on the South Side of Chicago, one resident, Hazel Johnson, started an organization called People for Community Recovery. Initially the group aimed to get public officials to confront problems such as leaking ceilings, low water pressure, and contaminated water.[18]

Soon Johnson significantly broadened her group's agenda, to encompass serious environmental concerns impacting the entire community. Her husband had died in 1969 from cancer, in his early forties. Johnson discovered high incidences of cancer throughout the area, including in babies and children. Since its creation after World War II, Johnson learned, the neighborhood had become a dumping ground for toxic chemicals. It was surrounded by fifty landfills, had soils infiltrated by sludge, and homes riddled with lead paint. Johnson began speaking out, contacting media, lobbying lawmakers, speaking at conferences. She led "toxic tours" of her neighborhood. These actions earned her the title "mother of the environmental justice movement."[19]

Journalists, including those in minority communities, began writing about the toxics poisoning neighborhoods, declaring the cause to be "institutional racism," which some called "ecoracism" due to constraints on "housing and mobility systems for minorities," zoning ordinances, deed restrictions, and "other land use mechanisms" that kept nonwhites in impoverished and segregated neighborhoods.[20]

In October 1991 activists from around the world gathered at the first National People of Color Environmental Leadership Summit in Washington DC. Sponsored by the United Church of Christ's Commission on Racial Justice, it aimed to bring the topic of environmental justice into broader public view.[21] The more than six hundred attendees included African Americans, Latinos/as, Asian Americans, and members of tribal nations, who experienced significantly higher incidents of miscarriage and cancer due to uranium mining and other kinds of environmental degradation on reservations. Many belonged to All Red Nations, a group created to challenge "ongoing attacks on Indian culture, health, and lands."[22]

Participants also included former residents of the Bikini Islands, who had been relocated after World War II, when the U.S. government began testing nuclear weapons on the site. Eventually the former residents tried to return home, but continuing high levels of radioactivity precluded it. The Bikini example reminded conference attendees that places where "other people" lived had for too long been considered inconsequential in the larger environmental movement. For too long, industry had been allowed to target "communities of color": "It's much more acceptable to dump this stuff in communities of color."[23]

Pamela Lee attended the summit and came away determined that the movement was "going to always fight for the protection of all—public health of all, the ecology of all." Representatives from mainstream groups attended as well, reflecting their belated recognition that the movement they had long dominated was being redefined and expanded. They needed to do some fence-mending. Since the late 1960s some establishment figures, including the former Sierra Club executive director David Brower, had touted zero population growth as an antidote to environmental destruction. Many people of color, and not just environmentalists, saw this focus as akin to the eugenics movement of the early twentieth century, an effort to limit the number of children born to minority and poor parents.[24]

Michael Fischer, the current Sierra Club executive director, spoke at the summit. He acknowledged that his organization had been "conspicuously missing from the battles for environmental justice." But it was "time to work and look to the future," he said. Establishment groups were "not the enemy." Many attendees remained skeptical. In her speech the summit leader and African American activist Dana Alston directed some of her remarks toward mainstream leaders. Environmental justice advocates did not want paternalism, she argued, but "equity, based on mutual respect, mutual interest, and justice."[25]

And they wanted middle-class, white environmentalists to own their privilege and stop promoting the notion, as Lee put it, that their communities experienced the same problems—toxic dumps, for example—as the nonprivileged. But summit leaders mischaracterized some aspects of the mainstream movement, which, they asserted, was still led by white men, who cared mostly about protecting wilderness. In fact the movement had moved far beyond wilderness protection since the early 1960s. And women had spearheaded many significant campaigns, though they were not women of color.[26]

Women took the lead in environmental justice campaigns as well. Having been leaders in their churches, schools, and neighborhoods, they understood "the ways in which the poor, people of color and other marginalized groups [bore] the brunt of environmental harm."[27] In the 1960s white feminists had coined the phrase "The personal is political," but for low-income and minority women, the words meant something different: their families were directly impacted by environmental degradation, so their campaigns were waged largely on behalf of children—theirs and other people's. They were engaged, as one writer noted, in "survival politics."[28]

The summit "galvanized" the environmental justice movement and gave activists a "new language to define their work."[29] California, long a center of environmental activism, soon became pivotal to the new, expanded movement as well. Leaders included Marie Harrison, an African American mother and grandmother living in the low-income neighborhood of Bayview–Hunters Point in San Francisco, whose residents faced a number of serious health problems as a result of industrial contamination from two power plants and a company that stored hazardous chemicals. While much of San Francisco enjoyed the effects of the bay cleanup and restoration by the 1990s, Bayview–Hunters Point remained riddled with pollutants. A century earlier it had been the site of a slaughterhouse. In the 1910s the federal government built the Hunters Point Naval Shipyard, one of the largest on the West Coast. After World War II it became the site of the Naval Radiological Defense Laboratory, which studied the effects of radiation and nuclear weapons and, in the process, contaminated the soil and air. In 1929 PG&E built the Hunters Point Power Plant, which burned fossil fuel, spewing toxic chemicals into the air. The Mirant Power Plant, built in 1940, was a natural gas and diesel–burning facility located in the nearby Potrero neighborhood. The area also housed the Bay Area Drum Company. Built in the 1940s, it stored chemicals—oils, solvents, paints, and asphalt products among them. The residue from these items saturated the ground and created air pollution.

The Hunters Point Power Plant was located directly across the street from an apartment complex where Harrison's grandson lived. A longtime community activist, she began to investigate the possible causes when her grandson became chronically ill. He "used to suffer tremendously," she told a reporter. "He had asthma attacks and chronic nosebleeds. He would wake up some mornings with his pillow soaked with blood." Soon she discovered that many others in the complex also had serious illnesses. These included emphysema, congestive heart failure, and several different kinds of cancer. When she began looking at the larger apartment complex, she found that many children were "sick with asthma or another pulmonary disease." In one unit "three families out of five ... had breathing problems": "As I went up [to higher floors], it got even worse. As you get to the top of the hill, you find babies who just got here having skin rashes and a hard time breathing. Babies! Good God almighty!"[30]

She began an intensive campaign: "I became the biggest advocate for shutting down this power plant. I told [officials], 'I have a vested interest in seeing this plant close. I will be after you. And if I lose, I'm coming back.'"[31] Harrison soon expanded her focus to other environmental hazards. "There are close to 200 leaking underground toxic sites in Bayview–Hunters Point," she said. "There are two superfund sites: the shipyard and the Bay Area

Drum Co. There is the . . . sewage treatment plant." And more distressing information was yet to come. As Harrison told another reporter, "I'm no science major, but I do have common sense. When I see feces floating in the water, orange foam bubbling at the surface, an exodus of fisherman, and notice that pigeons . . . won't even fly overhead anymore, I know something's gotta be wrong."[32] Most enraging to residents of Hunters Point, perhaps, was the recognition that the power plants provided electricity to all of San Francisco, but only their neighborhood and its adjacent areas had to live with dangerously high levels of toxic pollution.

Harrison understood that "a demoralized, poor, underrepresented and poorly educated community is the least equipped to fight for justice," so she wrote articles for the *San Francisco Bay View*, a neighborhood newspaper, and became an activist with Greenaction, a San Francisco–based, multiethnic organization formed in 1997 that works with low-income neighborhoods on environmental issues. Greenaction, like other organizations, relies heavily on media to raise public awareness of events such as marches, large protests, and appearances before numerous government bodies. Constant pressure from Greenaction and other community organizations nudged public officials toward the notion of shuttering the Hunters Point plant, but foot-dragging slowed the process. First one deadline passed and then another. In the early 2000s Harrison, along with other members of the community, and Greenaction blockaded the plant's gate. Finally, in 2008, the plant was closed. The Mirant plant closed in 2011.[33]

Cleanup on a variety of fronts is still occurring and local environmentalists are still keeping watch over the process. In February 2018 Harrison penned an op-ed for the *San Francisco Chronicle*, castigating the city for enabling the U.S. Navy to falsify data on cleanup of the Hunters Point shipyard so it could sell the land to developers for housing. A "century or more of soil and groundwater pollution . . . is not being cleaned up to the higher standards required for the new uses," she wrote. "The city is trying to double the population" of the area "while avoiding pollution cleanup." Therefore any new residents in the area will live under the same cloud of toxic pollution as those who have lived there for decades.[34] Harrison, a nonsmoker, died of chronic lung disease in 2019. She was seventy-one.

Four hundred miles south of San Francisco, the Mothers of East L.A. (MELA) emerged "from nowhere" in the mid-1980s to oppose a 1,700-acre proposed maximum security prison near Boyle Heights, part of a larger area known as East Los Angeles. Relatively small—just over twenty square miles—it is home to nearly 300,000 residents, virtually all of them Latino/a or African American, most with limited income and education. It is also the most heavily industrialized part of Los Angeles, with shipping, trucking, and oil-refining facilities and plastics, garment, and furniture manufacturers.[35] The state Department of Corrections had lobbied to locate the $100 million facility in Los Angeles County because so many inmates in the overall state prison population came from the region. Few proponents believed they would face any blowback from residents, even though the planned facility would be close to more than thirty schools. The state also had approved a $29 million waste incinerator to be constructed in the nearby, heavily industrialized town of Vernon. It would burn 225,000 tons of toxic waste per year. The region already had a publicly owned incinerator; the new one would be privately owned.

The more than four hundred MELA members may not have had much education or experience in political organizing, but they did have bitter experience with injustice that fueled their activism. In the 1950s many had lost their homes in the frenzy of freeway building that demolished whole neighborhoods. Juana Gutierrez had to move twice. "The people accepted it because the government ordered it," she said.[36] "I remember that I was angry and wanted the others to back me, but nobody else wanted to do anything." When her children came along, she turned her anger into advocacy on their behalf. "I started by joining the [PTA] and we worked on different problems," explained Gutierrez, who eventually became president of the local PTA. When she learned of drug dealers in parks, she "got the neighbors to have meetings . . . knock[ing] at the doors, house to house." Another MELA member, Erina Robles, also helped out at her children's school. "And before you knew it, we had a big group of mothers that were very involved."[37]

Fortunately for Gutierrez and others, the timing of the environmental justice movement in California coincided with increased numbers of non-Anglo politicians holding office, from the local level to state positions. African

Americans had created the Legislative Black Caucus in 1967, and Latinos/as followed with the Latino Caucus in 1973. It was not until the 1980s, however, that either group attained leadership positions or had enough clout to make a significant impact. In 1980 Willie Brown, a San Francisco Democrat, became the first African American elected by his peers as speaker of the state assembly. Cruz Bustamante, a Fresno-area Democrat, became the first Latino speaker in 1996. Some of the elected lawmakers were women: Yvonne Brathwaite Burke, an African American, and Gloria Molina and Lucille Roybal Allard, both Latinas. All three women were from Los Angeles.[38]

In fact it was a representative for Molina, then a Los Angeles City councilwoman, who informed the women of East Los Angeles about the proposed prison. She told them, "You know what is happening in your area? The governor wants to put a prison in Boyle Heights!" recalled Gutierrez. "So I called a Neighborhood Watch meeting at my house."[39] In addition to Molina, MELA members had help from the local priest, Father John Moretta of the Resurrection Catholic Church, who had been instrumental in creating and naming the organization. MELA members also included grandmothers and male family members. As Save the Bay had done decades earlier, MELA recruited men, though the women did most of the work. The men were as dedicated as the women, organizers insisted, but they had to work, often at more than one job, to support their families, and thus had limited time and energy.

When the Department of Corrections agreed to hold a public hearing on the prison in 1987, more than seven hundred Boyle Heights residents showed up. That summer more than two thousand people marched from Resurrection Church to a bridge connecting East Los Angeles and downtown Los Angeles. Marchers carried signs bearing the message "No Prison in ELA!" Marcher Aurelia Castillo explained, "You know if one of your children's safety is jeopardized, the mother turns into a lioness. We have to have a well-organized, strong group of mothers to protect the community and oppose things that are detrimental to us." MELA continued to oppose the prison. Donations from numerous supporters enabled members to travel to Sacramento by bus to testify against it. In 1991 state lawmakers voted against the facility, which was never built.[40]

Plans for the incinerator, slated for a nine-acre parcel near the interchange of two freeways—the Santa Ana and Long Beach—had been kept under wraps for two years when MELA and other community groups discovered them in 1987. By that time most of the permits for the facility, to be owned and operated by California Thermal Treatment Service, had already been granted. The South Coast Air Quality Management District had ruled in 1985 that no environmental impact report was needed for the project. State Assemblywoman, later Congresswoman Lucille Roybal-Allard complained that the public had not been properly notified, so a public hearing was scheduled for December 1987.

MELA members s attended the meeting, along with hundreds of others from all over East and South Los Angeles, including members of a "predominantly black community organization in South Central Los Angeles."[41] "Pushing baby strollers and wearing white kerchiefs, they are not exactly a formidable sight," wrote a *Los Angeles Times* reporter. "Yet the Mothers of East LA . . . is the vanguard of a fledgling environmental movement." He quoted Gutierrez: "We're ready to fight against all the injustice that they try to dump on our community." Added another member, Lucy Ramos, "There is a limit to how much junk they can dump on people, and this community has reached it."[42]

At the meeting, irate audience members shouted down speakers, leading officials to postpone any action on the incinerator for at least a month. Continuing community pressure led to a decision by the Air Quality Management District to require an environmental impact report for the project. Meanwhile, Roybal-Allard successfully introduced state legislation mandating such reports for all hazardous waste facilities. The Western Center on Poverty and Law filed two lawsuits against the proposed incinerator. By May 1991 California Thermal had abandoned plans for the incinerator. "There was a lot of political pressure that was brought on the district," said an attorney for the company.[43] As one MELA member put it, "You know, nothing ventured nothing lost. You should have seen how timid we were the first time we went to a public hearing. Now, forget it. I walk right up and make myself heard and that's what you have to do."[44]

It was the aftermath of a January 1994 earthquake that fueled the activism of Esperanza "Linda" Marquez. The 6.7-magnitude temblor struck before dawn. Centered in the city of Reseda, about twenty miles north and west of downtown Los Angeles, it killed sixty people and collapsed whole sections of major freeways, parking structures, and office buildings. In the aftermath, as the Department of Transportation worked to repair roads that carry millions daily through the region, a company called Aggregate Recycling won a contract to move and crush tons of destroyed concrete. As a storage site, Sam Chew, the owner of Aggregate, chose a 5.5-acre empty lot in the majority-Latino/a city of Huntington Park, in Southeastern Los Angeles County. The vacant lot sat near homes, schools, and businesses.

As trucks continued dumping the concrete, the pile of rubble grew to an estimated five stories tall. Residents named it "La Montana," a description not meant to inspire awe or admiration. The company's plan was to grind up the concrete, crush it, and sell it for building new roads. But Chew could not sell it as fast as he dumped it, so the pile kept growing. Meanwhile, day and night, trucks drove through the neighborhoods carrying and depositing the concrete. The noise and dust proved intolerable. One night the "dust was so intense that two neighborhood boys were taken away by ambulance, oxygen masks strapped to their faces." Other residents complained that the pervasive dust caused headaches, bloody noses, and constant shortness of breath. It coated their yards like snow, killing their lawns and gardens. Rick Loya, a Huntington Park City Council member, was hospitalized when one of his lungs nearly collapsed from the airborne pollution.[45]

Marquez, then in her seventies, quickly became involved in an effort to stop the dumping and crushing. A longtime activist, she knocked on hundreds of doors and invited fellow residents to a meeting at her home, where she spearheaded the creation of a group called La Causa. She and other residents also joined Communities for a Better Environment (CBE), a national organization that "combined scientific research and legal expertise to expose and fight industrial pollution in low-income communities of color."[46] Soon residents began filing complaints against Aggregate with the city of Los Angeles. In November 1994 they showed up at a Los Angeles Planning Commission meeting to discuss the renewal of Aggregate's conditional use permit to continue the dumping and crushing operation. Chew argued that Aggregate had taken steps to reduce the noise and dust. "We don't have those sounds anymore," he said. As for dust: "There's no problem there either. We keep the dust piles water[ed] down all day long." After the commission voted to renew Aggregate's permit, Marquez said, "I can't believe they'll just go on making noise.... It's really making our lives miserable."[47]

Huntington Park residents appealed the decision, and a month later more than 150, including many children, showed up to a Los Angeles City Council meeting to discuss the pending permit. They wore surgical masks and carried placards in Spanish and English reading "Let Us Breathe." Attorneys for Huntington Park businesses also complained about disruptions. Residents were hopeful, but skeptical, when council members ordered an air quality analysis to determine the source and extent of air contamination.

Within a year Aggregate had been ordered to stop hauling and crushing concrete, but the ruling did not mean immediate removal of La Montana. Chew declared bankruptcy, and it took until 2006, and numerous actions by CBE and La Causa, before the pile—weighing six hundred tons—was finally removed. By 2007 the vacant site had become part of the so-called Toxic Tour of Los Angeles, staged by CBE. Tourists traveled by bus through areas blighted by environmental destruction. Other toxic sites included a school where teachers and staff suffered higher than normal incidents of miscarriage. Today the site where La Montana once stood houses Linda Marquez High School, opened in 2012. Part of the Los Angeles Unified School District, it contains a School for Social Justice.[48]

The Asian Pacific Environmental Network (APEN) was created in 1993; it works on behalf of low-income Asian Americans in San Francisco's East Bay, lobbying government officials, promoting legislation, and overseeing projects to ensure economic and environmental equity. Peggy Saika, a longtime civil rights activist, served as its first executive director, from 1993 until 2000.

Saika was born in an Arizona internment camp near the end of World War II, a beginning that revealed to her "the fragility of democracy." She was reared in Sacramento; her father worked in the fields, and her mother—despite an allergy to peach fuzz—worked in area canneries. To put herself through California State University, Sacramento, Saika worked as a hairdresser, while also becoming involved in organizations that fought "for the rights of immigrants and refugees that were being exploited." Their agenda included establishing services for low-income seniors, prodding local authorities to improve schools, and promoting criminal justice reform. Her clients were Asian, African American, and Chicano/a. After graduation Saika moved to New York, where she volunteered for Women Against Rape. Back in the Bay Area in the 1980s, she went to work for the Asian Law Caucus, which initiated legal cases on behalf of low-income Asian American clients. "You build this thick strand of organization within one's own community because no one else is going to do it for you," she said.[49]

The 1991 National People of Color Environmental Leadership Summit was "a life-changing experience . . . deep and profound," for Saika. In its aftermath Asian Americans sought to create their own "presence within the environmental justice movement," one focused not just on "where we live, work and play but where we pray and go to school." The result was APEN, an organization "built through the lens and leadership of women."[50]

An early APEN campaign took place in the city of Richmond, which sat in the shadow of a massive Chevron oil refinery, as well as other sources of toxic pollution. Many residents were immigrants or refugees who spoke little or no English. To accommodate the large Laotian community, in 1995 APEN created the Laotian Organizing Project, centered on a seemingly unlikely group—teenaged girls—who could be trained as community activists. The girls, second-generation and bilingual, signed on for a five-year commitment. In its initial phase, the project aimed to enhance the teens' self-esteem and emphasized team-building and organizational skills. Participants attended workshops and conferences. Over time the girls learned to advocate on behalf of their family and community.

A test of the project's effectiveness came in March 1999, when an explosion and fire from the refinery sent hundreds of people to area hospitals with breathing problems. City and county officials had warned residents to remain indoors, but the warnings were issued only in English, significantly exacerbating health risks. The girls in the Laotian Organizing Project participated in a campaign urging officials to issue warnings in other languages as well. They attended press conferences, holding aloft signs reading, "Safe systems for all people." Warnings in multiple languages began in 2001. They found less success in an effort to defeat Proposition 227, a 1998 California ballot initiative limiting the amount of time students not proficient in

English could stay in bilingual classes. The measure reflected anger among many California voters at the time, who resented money spent on bilingual students, believed schools kept limited-English students in separate classes too long, and promoted moving them quickly into English-only classes. The proposition passed by a margin of 61 to 39. It was repealed in 2016.[51]

Not all of those experiencing pervasive toxic pollution lived in minority communities. Residents in the small, rural, working-class town of Glen Avon, fifty miles east of Los Angeles in Riverside County, lived next to the Stringfellow Rock Quarry. In 1956, with approval from the state and Riverside County officials, the seventeen-acre parcel was turned into a dump site, called the Stringfellow Acid Pits. Neighbors won assurances from local and state officials that the solid bedrock would prevent any seepage. Over the next sixteen years, more than two hundred industries dumped toxic waste in the pits. The U.S. Air Force dumped chemicals used for refurbishing missiles. Private companies dumped waste products as well; they included General Electric, McDonnell Douglas, Northrup, Montrose Chemical Co., and Rockwell International. Altogether, a total of thirty-four million gallons of waste was dumped.

Residents may have had concerns, but they remained mostly silent. Then came a series of pounding rainstorms. The first one, in 1969, caused the pits to overflow, spewing toxics across the area. By the 1970s it had become clear that the waste products were leaking into the area's groundwater. Authorities shut down Stringfellow in 1972, but the damage had been done. In 1978 another series of storms struck, leading officials to release water from a nearby dam, sending 5.5 million gallons of waste pouring into backyards, schoolyards, and wells. "The toxic brew ... caused sneakers and blue jeans to disintegrate" and led Penny Newman to become a community activist.[52]

Newman was in her forties, married, with two sons, and had grown up in the area. A special education teacher, she hated speaking in public, but silence was not an option after her sons began to have health problems—difficulty breathing, dizzy spells, headaches, and blurred vision. She herself developed benign tumors and recurring tremors. "You don't have to be an elected official or an industry executive to have an impact on waste policy," she decided.[53] In 1980 Congress set up the Superfund program and gave the Environmental Protection Agency the authority to order the cleanup of contaminated areas. The disaster at Stringfellow put it on the EPA's first list of sites, but no action was immediately forthcoming. Who was to blame? State and local officials had failed to address the problems when they first emerged. Polluters and insurance companies fought each other over liability.

Newman organized other parents to push for immediate cleanup and to lobby California officials to follow suit. She pored over technical reports and soon became an authority on the dump site. She met with government and industry officials. When cleanup started, it proceeded at an excruciatingly slow pace. Predictably, disagreements emerged among the parties involved. When Newman learned that the EPA had threatened to cut funding for a staffer working on the cleanup, she organized a fundraiser on his behalf.

In 1984, "after a plume of toxic underground water was discovered seeping toward Glen Avon's water wells," the town's 3,800 residents filed a lawsuit to compel industries and the state to speed the cleanup. By the late 1980s Newman had left teaching and become a full-time environmental activist, working as the western director for Citizens Clearinghouse for Hazardous Waste, an organization that educates members of the public on pollution. She traveled the country speaking with other activists, including Lois Gibbs, leader of the Love Canal Homeowners Association in Niagara Falls, site of a seventy-acre hazardous waste dump. She participated in national gatherings, such as the Women in Toxics Organizing conference in Arlington, Virginia. Newman and her colleagues in the western delegation arrived at the 1987 gathering wearing "bright new shocking-pink tee shirts." The front bore a stenciled figure of a woman flexing her muscles and the words "Tough Women Against Toxics." The back read, "WE CAN DO IT."[54]

The legal battle took decades to settle. In 1992 private companies involved in dumping eventually agreed to pay $150 million to help government agencies expedite a cleanup, though the actual estimated cost was as much as $750 million. "It's just kind of another step in a long journey," Newman said of the legal action. "Maybe instead of them sitting in their offices and arguing who pays for what, now they'll have time to actually get out to the site and start doing something."[55]

The following January, trial began in a second lawsuit—the largest civil suit in the nation's history to date—seeking monetary damages for area residents from government agencies and the companies responsible for Stringfellow. State officials suggested that eradicating all remnants of the toxins could take as long as 446 years. In addition to health problems, residents who tried to sell their homes found scant interest from buyers who were understandably reluctant to live near the remains of a toxic dump. "We understand we're at the cutting edge, not only within the legal system—which hasn't had a long time dealing with toxic tort cases—but within the medical field too," Newman told a reporter. "The medical profession is way behind the ball in exploring the health effects on people exposed to toxic chemicals. Few physicians are trained in that field, and medical schools haven't yet focused on it." Since the trial had so many plaintiffs and defendants and was expected to last up to nine months, two thousand prospective jurors had been summoned.[56]

By the time the plaintiffs' case ended in November 1994, five hundred lawyers had worked on it. Residents won a total of $109 million from the state, Riverside County, the polluting industries, and the owner of the dump site. They had paid out more than $30 million in legal costs.[57] Cleanup of Stringfellow continues. Now retired, until recently Penny Newman served on the board of the Center for Community Action and Environmental Justice, an organization she founded in the late 1970s, devoted to "empowering diverse communities to create safer, healthier, toxic-free places to live, work, learn and play."[58]

By the end of the twentieth century, the environmental justice movement could count significant victories. It had become a global phenomenon as it framed a new narrative on the relationship between poverty, racism, and environmental destruction. In 1994 President Bill Clinton signed Executive Order 12898, which directed federal agencies to "identify and address the disproportionately high and diverse human health or environmental effects of their actions on minority and low-income populations, to the greatest extent practicable and permitted by law." It also required heads of agencies to develop strategies for implementing environmental justice.

The movement has continued to gain momentum in the twenty-first century, as women found recognition for their leadership. In 2002 the People of

Color Environmental Justice Leadership summit honored the contributions of women, who are "less vested in the status quo" and "unwilling to believe empty promises" than their male counterparts. As community caretakers, they are "first to notice what is amiss in the family, community, and local environment."[59]

Media attention fueled wider public interest in the movement. "Hardly a day passes without the media discovering some community or neighborhood fighting a landfill, incinerator, chemical plant, or some other polluting industry," one activist wrote in 2014. "This was not always the case."[60] Celebrity activists raised its profile even higher. Erin Brockovich was a single mother and legal assistant in a Los Angeles–area law firm when she began working on a case involving PG&E and the small, impoverished Mojave Desert town of Hinkley, California. The utility operated a compressor station in Hinkley, which transmitted natural gas along a pipeline. It used water-cooled towers for the compressed gas. For fourteen years, beginning in the 1950s, PG&E stored the cooling water—more than 300 million gallons—in nearby unlined ponds. Unbeknownst to residents, the water contained chromium, a carcinogen, which seeped into groundwater and ultimately infiltrated the water in local wells. After PG&E finally notified residents in 1987, they filed a class action suit. Brockovich began investigating reputed cancer clusters in the region and urged reluctant residents to join the lawsuit. In 1996 PG&E settled for $333 million. It was a story made for Hollywood, particularly since the slender, blonde Brockovich looked like a movie star. In fact Julia Roberts played her in the film and earned an Oscar for her performance.[61]

The movement also compelled mainstream organizations to enlarge their own agendas. For example, the Sierra Club created a National Environmental Justice Program. Leslie Fields, a director of the program, called environmental justice "the conscience of the environmental movement."[62] As more people of color ran for and won political office, they sponsored and enacted legislation. Connie Leyva was born and reared in Chino, California, and began working in a local grocery store while still in high school. In her twenties she became a representative and later president of the United Food and Commercial Workers union. In 2004 Leyva was elected the first female president of the California Labor Federation. A decade later she won election to the California State Senate, where she authored several environmental justice bills, among them bills to require all children enrolled in Medi-Cal to be tested for lead poisoning; to ensure that local governments include environmental justice elements in their general plans; and to maximize climate investments in low-income communities.[63] And in November 2019, for the first time, six Democratic presidential candidates participated in a forum on environmental justice in South Carolina.[64]

In 2014 longtime leaders came together to discuss what they viewed as the environmental justice movement's wider impact on society and politics. They could point to many victories. In 1994, for example, only four states had enacted environmental justice laws, but twenty years later all fifty states had enacted them. Three hundred grassroots groups had existed in 1992; twenty-two years later the number had grown tenfold. Dozens of books had been written, and colleges included the topic in courses and curricula. The movement had "profoundly . . . transformed the face of U.S. environmentalism." It had "paved the way for justice," adherents noted, "but didn't guarantee it." For example, Clinton's 1994 executive order had addressed environmental inequities but included the proviso that official action could occur "to the greatest extent practicable," not exactly a promise of action. And despite the gains, some communities continued to be "unnecessarily exposed to toxic pollution." The "most potent predictor of health" remained the zip code in which people resided.[65]

The port city of Wilmington, south of Los Angeles and west of Long Beach, is one of those unfortunate zip codes. Nearly 90 percent Latino/a, it is home to the third largest oil field in the United States. Residents there have experienced lung diseases and cancers as the result of living with the constant pollution from hundreds of oil wells, as well as from diesel-burning ships in the harbor. Meanwhile, a few miles away, the affluent city of Palos Verdes does not allow billboards to ruin the view. In 2006 California enacted tough global warming legislation, but four years later oil companies raised millions of dollars for Proposition 23, an industry-backed ballot initiative to overturn the earlier law. Backers of the proposition argued that it would cost workers jobs.

Alicia Rivera had fled El Salvador's violence in 1980. In 2010 she was an activist with Communities for a Better Environment in Southern California.

The *Los Angeles Times* columnist Steve Lopez spent a day with Rivera as she went door to door handing out anti–Proposition 23 flyers. She was, he said admiringly, "a detective, a rabble-rouser, a crusader." At one home a seriously ill resident said she hoped she would live long enough to vote.[66] In the end, Proposition 23 was defeated, but although the battle was won, the fight continues in a political climate that is not always supportive of environmental activism, and at a time when some—both in and out of government—would prefer to move the clock backward and to ignore serious and looming environmental crises.

CONCLUSION

When Rachel Carson published her seminal book, *Silent Spring*, in 1962, she hoped to raise the alarm about the dangerous and unregulated use of lethal pesticides. It probably is safe to say that neither she or anyone else could have envisioned the emergence of a mass environmental movement that went far beyond challenging pesticide use, to confront virtually every aspect of American society and culture. Yet when Carson died in 1964, this new movement was already in the process of being born—in San Francisco, on California's Central Coast, in Los Angeles, and elsewhere. Its focus already extended to air and water pollution, lack of open space, and destruction of natural resources through overdevelopment.

No one, it is safe to say, could have foreseen that such a movement would be led—at least in its first iteration—by white, upper-middle-class housewives. Before they began their journey, surely the women themselves could not have imagined such an unlikely scenario. They lived in affluent communities: Berkeley, California's Central Coast, Malibu, even Beverly Hills. They had well-appointed homes, household help, wide circles of friends. Few worked outside the home, though they participated in many volunteer activities, which instilled a sense of competence and confidence. But they had virtually no involvement in electoral politics, except perhaps as voters.

Yet their communities, however affluent, did not isolate them from problems created by the post–World War II obsession with growth and development. The San Francisco Bay was filled with garbage, and industry owned much of the land around the bay. Oil companies owned land on the pristine sand dunes in southern San Luis Obispo County, where the giant utility PG&E planned to build a nuclear power plant. Then there was Los Angeles, ground zero for developers. The basin's air was barely breathable for much of the year because of smog that made it impossible to see the nearby mountains. And land that once held thousands of acres of orchards and agriculture had been turned into subdivisions. Building on the sparsely populated hillsides caused floods and fires. With no one—apparently—willing to address serious issues or problems, the housewives decided to do it themselves. They did not start out to challenge the status quo, only to find solutions to individual problems. But within a decade they had laid the foundation for the modern environmental movement.

Their journey unfolded slowly. At first they gathered around dining-room tables, pondering how to get elected officials to take smog seriously or how to get cities to address pollution in the bay. They met in front yards, where they watched bulldozers lopping off hillsides for planned subdivisions. When they contacted local officials to get answers or established conservation organizations to get ideas, they were met by silence or (metaphorically) patted on the head. They quickly learned how politics worked: public officials received campaign donations from developers, thus they had no incentive to slow down or stop building. And established organizations like the Sierra Club had their own agendas and were not inclined to pivot to issues outside their areas of expertise or interest.

So the women used what they had: friends with money, and knowledge of how to organize from their experience with volunteerism. For the former, they utilized Christmas card lists, the names of acquaintances from volunteer organizations. They wrote up flyers and laboriously listed the names of donors on three-by-five cards. Donors provided names of others, and soon they had hundreds of participants—most of them women—involved. If, like Kathleen Goddard Jones or Jill Swift, they belonged to the Sierra Club, they led hikes on the Nipomo Dunes or in the Santa Monica Mountains, stopping occasionally to describe the landscape and solicit more money and volunteers.

They also utilized public relations strategies they had learned through other organizations: to get attention, you have to do something bold. So members of Stamp Out Smog attended meetings wearing gas masks; they made a smog birthday cake and took it to public meetings. Members of Save the San Francisco Bay sent bags of sand and crafted messages designed to embarrass politicians they feared would oppose their efforts for a state agency to oversee development in the bay.

This was a decidedly different strategy than that favored by established organizations, which utilized a more formal approach, with boards of directors and layers of decision-makers. But the women's strategies worked. Over time, public officials learned that they ignored the women at their peril. For their part, the activists grew more comfortable speaking in front of large groups, even as they labored to appear traditional in dress and mannerisms.

By the early 1970s the housewives who had begun as nervous neophytes had become battle-hardened political experts. They had slowed or stopped the dumping in San Francisco Bay, convinced PG&E to abandon plans for a power plant on the Dunes, and forced Los Angeles city and county officials to push strong smog-control measures. And they were on the way to garnering congressional support for a vast recreation area in the Santa Monica Mountains. They also proved instrumental in the passage of environmental legislation, including the California Coastal Act in 1972, which led to the creation of the California Coastal Commission, the agency that controls development along the state's 1,100-mile coastline. By the mid-1970s some of the activist women had moved into politics themselves. Claire Dedrick, for example, a member of Save the San Francisco Bay and many other organizations, joined the administration of Governor Edmund G. "Jerry" Brown Jr. in 1975, serving as resources secretary. She later became the first women to serve on the state Public Utilities Commission.

With all of their successes, the women failed to address, or ignored, serious problems impacting communities unlike their own. Saving the Santa Monica

Mountains from developers, for example, meant keeping out condominiums and apartments for lower-income residents. Saving the San Francisco Bay did not extend to addressing the destructive impacts of oil refineries, toxic chemicals, and industrial waste on poor and minority communities. It would be left to residents of these communities—with far fewer resources and virtually no access to the political system—to craft their own movement, which they called "environmental justice." Women led many of these campaigns as well. In East Los Angeles they stopped construction of a prison that was to be built near schools. They forced a contractor to remove a five-story-high mountain of concrete. In San Francisco they forced the closing of Bayview–Hunters Point shipyard. In the East Bay they compelled Richmond politicians to issue evacuation orders in languages other than English.

Like the women before them, environmental justice activists succeeded in pushing the movement in ever expanding directions, forcing it to grapple with issues such as the relationship of poverty and racism to environmental destruction. In the early 1990s Sierra Club leaders—who had broadened their own agenda in response to the earlier generation of activists—acknowledged that their organization had been "conspicuously absent" from the fight for environmental justice. Other organizations followed suit, and in 2018 a coalition of environmental groups created the Equitable and Just National Climate Platform to "guide, aid, and inform local, regional, and national policymakers, business leaders and civil society advocates."[1] State and national politicians have been forced to grapple with inequities as well. In 1994 President Bill Clinton acknowledged the movement with Executive Order 12898, mandating that all federal agencies examine policies to determine their impacts on the health and environments of low-income and minority populations. And in late 2019 six Democratic presidential candidates met in South Carolina to discuss environmental justice issues.

With generations of women working on behalf of environmental change, it seems appropriate to return to the woman often credited with starting it all. Rachel Carson may not have foreseen all the paths her work would take, but she might have anticipated *Silent Spring*'s profound impact on American science and politics. She would not have been surprised, however, at the controversy still surrounding the effort to save the planet from disaster. As she warned readers:

> We stand now where two roads diverge. But unlike the roads in Robert Frost's familiar poem, they are not equally fair. The road we have long been traveling is deceptively easy, a smooth superhighway on which we progress with great speed, but at its end lies disaster. The other fork in the road—the one "less traveled by"—offers our last, our only chance to reach a destination that assures the preservation of our earth.
>
> The choice, after all, is ours to make.... We should no longer accept the counsel of those who tell us that we must fill our world with poisonous chemicals; we should look about and see what other course is open to us.[2]

NOTES

INTRODUCTION

1. Stevens, "Feminizing the Urban West," 134.
2. Goddard Jones Papers, box 1, folder 1.
3. Stevens, "Feminizing the Urban West," 149.
4. Sellers, *Crabgrass Crucible*, 198.

1. "FEMININE WARRIORS"

1. "Women's Group Joins Fight on Air Pollution," *Los Angeles Times*, January 8, 1959, B1.
2. Before his election to the Los Angeles County Board of Supervisors in 1956, Dorn served as mayor of Pasadena, a city that suffered significantly from air pollution. As a supervisor, he considered his fight against smog to be his most significant accomplishment. However, his activism had a personal aspect: he was asthmatic. He served until 1972. During his sixteen years on the board, he was instrumental in the passage of more than one hundred antipollution laws.
3. "No Mystery, Same Smog, Dorn Avers," *Los Angeles Times*, November 27, 1958, A1.
4. Dorothy Townsend, "Anti-Smog Group Set to Storm Supervisors," *Los Angeles Times*, May 21, 1959, A1; "Women Unite to Begin War against Smog," *Los Angeles Times*, October 6, 1959, B1.

5. "On a Clear Day: 21st Anniversary of Smog Marked," *Los Angeles Times*, November 17, 1964, A1. Most newspapers still identified women by their husband's name in the mid-1960s, though that would soon change. They identified Afton Slade by her own name, possibly because she was divorced.

6. Jacobs and Kelley, *Smogtown*, 46.

7. Lynn, *Progressive Women*, 11–12.

8. Starr, *Golden Dreams*, x–xi.

9. Jackson, *Crabgrass Frontier*, 265.

10. Starr, *Golden Dreams*, 5.

11. Matthews, *Silicon Valley*, 18.

12. Jackson, *Crabgrass Frontier*, 267.

13. Waldie, *Holy Land*, 8.

14. Didion, *Play It as It Lays*, 15.

15. Napoli, *Ray and Joan*. The McDonald brothers had failed at many enterprises when they decided to open their hamburger stand in the 1940s. Initially they used carhops to carry ordered food to customers' automobiles. But eventually they scrapped that idea and decided to create an assembly-line model. Customers came into their restaurant, and the McDonalds sought to fill their orders within one minute. To do so, they bought eight blenders to make milkshakes and put some premade shakes in the refrigerator to ensure there was always a supply. They also mass-produced burgers and fries. By 1954 they had expanded throughout Southern California. They then partnered with Ray Kroc, who had a much more expansive view of the potential. In 1961 they sold out to Kroc, who took the business national, and then worldwide.

16. Gabler, *Walt Disney*, 480–530. Gabler traces Disney's life and discusses his obsession with creating an amusement park. The Los Angeles Dodgers and San Francisco Giants both arrived in 1958 from New York. The Anaheim-based Angels were created in the early 1960s, the San Diego Padres a few years later. The Oakland A's moved to California from Kansas City in the late 1960s.

17. Sellers, *Crabgrass Crucible*, 167.

18. The Sierra Club was created in California in 1892; Save the Redwoods (formerly the Sempervirens Club) was created in 1918.

19. Riley, *Women and Nature*, 17.

20. Bonta, *American Women Afield*, 237.

21. Carson, *Silent Spring*, 5, 22–23, 32.

22. Maril Hazlett, "Voices from the Spring: Silent Spring and the Ecological Turn in American Health," in Scharff, *Seeing Nature through Gender*, 112–13.

23. Nancy C. Unger, "Women, Sexuality, and Environmental Justice in American History," in Stein, *New Perspectives on Environmental Justice*, 54–55.

24. Douglas and Rothchild, *Voice of the River*, 228.
25. Douglas and Rothchild, *Voice of the River*, 222–28.
26. Alexiou, *Jane Jacobs, Urban Visionary*, 51–57.
27. Murie, *Two in the Far North*, 243.
28. Richard Sandomir, "Katie Lee, Folksinger Who Fought to Protect a Canyon, Dies at 98," *New York Times*, November 10, 2017, B7.
29. Swerdlow, *Women Strike for Peace*, 73.
30. Goodman and Dawson, *Mary Austin and the American West*, x, xi.
31. Austin, *Land of Little Rain*, 103.
32. Quoted in Goodman and Dawson, *Mary Austin and the American West*, 12.
33. Austin, *Land of Little Rain*, 59.
34. Hoffman, "Mary Austin, Stafford Austin, and the Owens Valley," 305.
35. Hoffman, "Mary Austin, Stafford Austin, and the Owens Valley," 315; Goodman and Dawson, *Mary Austin and the American West*, 76.
36. Schrepfer, *Nature's Altars*, 52.
37. David R. Lindberg, "Annie Montague Alexander: Benefactress of UCMP," UC Museum of Paleontology, accessed June 23, 2020, www.ucmp.berkeley.edu/history/alexander.html.
38. Smith, *Pacific Visions*, 116–18.
39. McClain, *Ellen Browning Scripps*, 9–71, 190–94.
40. Bonta, *American Women Afield*, 88–89.
41. Smith, *Pacific Visions*, 166–70. Information on Eastwood's involvement in Golden Gate Park is from Schrepfer, *Nature's Altars*, 87.
42. Evans, "Sixty Years with the Sierra Club," 2–6.
43. Bade, "Recollections of William F. Bade and the Early Sierra Club," 3–6.
44. Cameron Binkley, "Saving Redwoods: Clubwomen and Conservation, 1900–1925," in Cherny, Irwin, and Wilson, *California Women and Politics*, 168.
45. Unger, "Women, Sexuality, and Environmental Justice in American History," 52.
46. Binkley, "Saving Redwoods," 152.
47. Merchant, *Earthcare*, 57.
48. Binkley, "Saving Redwoods," 153; Burdette quoted in Unger, *Beyond Nature's Housekeepers*, 83.
49. Binkley, "Saving Redwoods," 157–60.
50. Binkley, "Saving Redwoods," 155–61.
51. Merchant, *Earthcare*, 133.
52. Schrepfer, *Nature's Altars*, 104.
53. "Rites for Girl Leader," *Los Angeles Times*, June 12, 1928, 8.
54. Farquhar, "Pioneer Woman Rock Climber," 2.
55. Farquhar, "Pioneer Woman Rock Climber," 27, 10.

56. Farquhar, "Pioneer Woman Rock Climber," 10.

57. Farquhar, "Pioneer Woman Rock Climber," 18, 51, 33.

58. Long in the planning, construction began in 1933 on the 1.7-mile suspension Golden Gate Bridge that links San Francisco and Marin County. Its completion in May 1937 was heralded by a week-long celebration. The day prior to its opening to traffic, 200,000 people walked the length of the bridge.

59. Walker, *Country in the City*, 88.

60. Sally Hauser and Joan Brown, "Oral History of Mary Summers," sound recording, Marin County Free Library, February 26, 1988.

61. Walker, *Country in the City*, 89.

62. Lee and Yung, *Angel Island*, 31–51. Livermore's son Norman became one of California's premier environmentalists. He served in the gubernatorial administration of Ronald Reagan (1967–75).

63. Quoted in Jim Wood, "Seven Decades of Activism," *Marin Magazine*, August 27, 2008, www.marinmagazine.com/August-2006/Seven-Decades-of-Activism/.

64. The city and county of San Francisco are a single entity.

65. Thiessen, *Dorothy Erskine*, 39.

66. John Jacobs, "An Interview with Dorothy Erskine: A Founder of SPUR and Greenbelt Alliance," 1971, reprinted SPUR, January 1, 1999, 1.

67. Jacobs, "An Interview with Dorothy Erskine."

68. Jacobs, "An Interview with Dorothy Erskine"; Walker, *Country in the City*, 136.

69. John B. Oakes, "Conservation: A Growing Force," *New York Times*, November 13, 1955, X33.

70. Hurley, *Environmental Inequalities*, 9, 47–68.

71. Julie Rynerson Rock commented on her mother's activism following Jackie Rynerson's death in 2017 at the age of eighty-seven in "Lakewood Mayor Kept City on Course," *Press-Telegram*, September 1, 2017, https://www.presstelegram.com/2008/04/16/lakewood-mayor-kept-city-on-course/. Rock has followed in her mother's footsteps, becoming an urban planner in San Bernardino County.

72. Meyerowitz, "Beyond the Feminine Mystique," 1467.

73. Wellock, *Critical Masses*, 51.

74. Carl Nolte, "Jean Kortum: Activist, Major Campaigner in San Francisco's 'Freeway Revolt,'" *San Francisco Chronicle*, October 10, 2007, B9.

75. Chris Carlsson, "Revisiting the San Francisco Freeway Revolt," *StreetsBlogSF*, June 11, 2009, https://streetsblog.org.

76. Malvina Reynolds, "The Cement Octopus," Schroeder Music Company, 1964. Reynolds's best-known song is "Little Boxes," used as the theme song for the television series *Weeds*, starring Mary Louise Parker, who becomes a drug courier to pay her bills after the death of her husband.

77. Wellock, *Critical Masses*, 49–57.
78. Johnson, "Captain Blake versus the Highwaymen," 75.
79. Tom Bott, "Lucretia Edwards 1916–2005," *Berkeley Daily Planet*, October 18, 2005, 1.
80. John Geluardi, "Open Space Advocate Honored with a Park," *Berkeley Daily Planet*, January 2, 2004, www.berkeleydailyplanet.com/issue/2004-01-02/.
81. Geluardi, "Open Space Advocate Honored with a Park."
82. Green, "Environmental Activism in Los Angeles."
83. Green, "Environmental Activism in Los Angeles." Hyperion contracted with Los Angeles early in the twentieth century. Few problems emerged until after World War II, during the period of massive growth. By 1957 it became impossible to treat all of the waste, so Hyperion began to release sewage five miles offshore. The company agreed to measures designed to mitigate continuing problems, but Los Angeles officials sought more time to ensure compliance. Heal the Bay emerged during a legal standoff between the city and the federal government.
84. Green, "Environmental Activism in Los Angeles."
85. Green, "Environmental Activism in Los Angeles."
86. Elaine Woo, "Environmentalist Began Heal the Bay," *Los Angeles Times*, October 14, 2008, A1.
87. Green, "Environmental Activism in Los Angeles."

2. SAVING THE SAN FRANCISCO BAY

1. McLaughlin, "Citizen Activist for the Environment," 60–70.
2. Petris, "Oral History Interview with Nicholas Petris," 287.
3. McLaughlin, "Citizen Activist for the Environment," 8–15.
4. McLaughlin, "Citizen Activist for the Environment," 28–40. The median age for women to marry was 21 in 1940. By 1950 it would drop to 20.3. McLaughlin said she heard stories throughout her childhood about her Cherokee great-grandmother.
5. McLaughlin, "Citizen Activist for the Environment," 56–60.
6. Quoted in Odell, *The Saving of San Francisco Bay*, 10.
7. McLaughlin, "Citizen Activist for the Environment," 93.
8. Harold Gilliam, "How the Bay Was Saved: Development Threatened to Fill It In," *SF Gate*, April 2007.
9. Daryl E. Lembke, "San Francisco Bay Endangered by Feverish Developing Pollution," *Los Angeles Times*, December 16, 1963, C1.
10. Ronald J. Ostrow, "S.F. Opens Its Golden Gate to Negativity," *Los Angeles Times*, February 18, 1990, A1.
11. Odell, *The Saving of San Francisco Bay*, 10–12.
12. McLaughlin, "Citizen Activist for the Environment," 97.

13. Quoted in Lembke, "San Francisco Bay Endangered by Feverish Developing Pollution."
14. McLaughlin, "Citizen Activist for the Environment," 98.
15. Gulick, Kerr, and McLaughlin, "Saving the San Francisco Bay," 1.
16. Gulick, Kerr, and McLaughlin, "Saving the San Francisco Bay," 85–86.
17. Gulick, Kerr, and McLaughlin, "Save the San Francisco Bay Association," v.
18. Kittredge, "Volunteers and Employment Careers," 2.
19. McLaughlin, "Citizen Activist for the Environment," 98–99.
20. Thiessen, *Dorothy Erskine*, 119.
21. Odell, *The Saving of San Francisco Bay*, xi.
22. Gilliam, "How the Bay Was Saved."
23. Bodovitz, "Joe Bodovitz," 20.
24. McLaughlin, "Citizen Activist for the Environment," 107.
25. Gilliam, "How the Bay Was Saved."
26. Bodovitz, "Joe Bodovitz," 20.
27. Odell, *The Saving of San Francisco Bay*, 13.
28. McLaughlin, "Citizen Activist for the Environment," 107, 105.
29. Quoted in Odell, *The Saving of San Francisco Bay*, 15, 103.
30. Myron Rumford was elected to the California state legislature in the 1940s. As a lawmaker, he is best known for the Fair Housing Act of 1963 that outlawed segregation in housing. A ballot initiative the next year overturned the measure, but the California Supreme Court declared the initiative unconstitutional.
31. Gulick, Kerr, and McLaughlin, "Save the San Francisco Bay Association," 135–37.
32. Kittredge, "Volunteers and Employment Careers," 11.
33. Lembke, "San Francisco Bay Endangered by Feverish Developing, Pollution."
34. McLaughlin, "Citizen Activist for the Environment," 137–38, 123.
35. McLaughlin, "Citizen Activist for the Environment," 137–38, 123–24.
36. Gulick, Kerr, and McLaughlin, "Save the San Francisco Bay Association," 7.
37. Walker, *Country in the City*, 115.
38. Scott, *The Future of San Francisco Bay*, 6–11.
39. Scott, *The Future of San Francisco Bay*, 20.
40. Scott, *The Future of San Francisco Bay*, 21.
41. Petris, "Oral History Interview," 246, 107.
42. Petris, "Oral History Interview," 253.
43. Odell, *The Saving of San Francisco Bay*, 18.
44. Bodovitz, "Joe Bodovitz," 24.
45. Odell, *The Saving of San Francisco Bay*, 18.
46. Odell, *The Saving of San Francisco Bay*, 19.
47. Gulick, Kerr, and McLaughlin, "Save the San Francisco Bay Association," 29, 23.

48. Gulick, Kerr, and McLaughlin, "Save the San Francisco Bay Association," 128.

49. Gulick, Kerr, and McLaughlin, "Save the San Francisco Bay Association," 33.

50. Bodovitz, "Joe Bodovitz," 31.

51. Kittredge, "Volunteers and Employment Careers," 3–6.

52. Lawrence E. Davies, "Steps to Save Bay Backed on Coast," *New York Times*, January 24, 1965, B4.

53. Odell, *The Saving of San Francisco Bay*, 23.

54. Odell, *The Saving of San Francisco Bay*, 13.

55. Odell, *The Saving of San Francisco Bay*, 27, 30. The jingle parodied a popular television advertisement for Pepsodent toothpaste: "You'll wonder where the yellow went, if you brush your teeth with Pepsodent."

56. Petris, "Oral History Interview," 247–50, 287.

57. "The Great Race of '61: Don Sherwood vs. Jim Lange," *Marin Nostalgia*, accessed June 24, 2020, http://www.marinnostalgia.org/portfolio/the-great-race-of-61-don-sherwood-vs-jim-lange/.

58. Gulick, Kerr, and McLaughlin, "Save the San Francisco Bay Association," 43.

59. Gilliam, "How the Bay Was Saved."

60. Bodovitz, "Joe Bodovitz," 41.

61. McLaughlin, "Citizen Activist for the Environment," 130.

62. Gilliam, "How the Bay Was Saved."

63. Greg Lucas, "Claire Dedrick—Scientist, Served on PUC," *San Francisco Chronicle*, April 9, 2005, B4.

64. Walker, *Country in the City*, 115.

65. Daryl E. Lembke, "Drive Pushed to Save S.F. Bay," *Los Angeles Times*, January 5, 1969, B1.

66. Gulick, Kerr, and McLaughlin, "Saving San Francisco Bay."

67. Ross MacDonald and Robert Easton, "Santa Barbarans Cite an 11th Commandment: 'Thou Shalt Not Abuse the Earth,'" *New York Times*, October 12, 1969, SM32.

68. Gulick, Kerr, and McLaughlin, "Save the San Francisco Bay Association," 44.

69. Gulick, Kerr, and McLaughlin, "Save the San Francisco Bay Association," 47–49, 157.

70. Dolezel and Warren, "A Case Study in Environmental Legislation," 356.

71. Dolezel and Warren, "A Case Study in Environmental Legislation," 356.

72. Dolezel and Warren, "A Case Study in Environmental Legislation," 356.

73. Petris, "Oral History Interview," 256.

74. Dolezel and Warren, "A Case Study in Environmental Legislation," 356, 361.

75. Dolezel and Warren, "A Case Study in Environmental Legislation," 364.

76. Gulick, Kerr, and McLaughlin, "Save the San Francisco Bay Association," 43.

77. Burt Folkart, "Ex-Park Systems Chief William Penn Mott Dies," *Los Angeles Times*, September 23, 1992, A3. Mott later became head of the National Park System.

78. Gilliam, "How the Bay Was Saved."
79. Jerry Brown named Rose Elizabeth Bird secretary of agriculture, and then chief justice of the California Supreme Court, making her the first female cabinet member in California history.
80. Peter Fimrite, "Sylvia McLaughlin, Co-founder of Save the Bay, Dies at 99," *San Francisco Chronicle*, January 21, 2016, B1.
81. Scott, *San Francisco Bay Area*, 316–17.
82. Gulick, Kerr, and McLaughlin, "Save the San Francisco Bay Association," i.
83. Gulick, Kerr, and McLaughlin, "Save the San Francisco Bay Association," 5.
84. Ostrow, "S.F. Opens Its Golden Gate to Negativity."
85. McLaughlin, "Citizen Activist for the Environment," 78–79.
86. McLaughlin, "Citizen Activist for the Environment," 141–42.
87. Gilliam, "How the Bay Was Saved." Gilliam died in 2016 at the age of ninety-eight. He was the last surviving member of the group that attended the 1961 meeting that formed Save the San Francisco Bay. He is considered the father of environmental journalism by many in the Bay Area. The author of thirteen books, he covered the environmental movement from 1961 until his retirement in 1995.

3. THE DUNE LADY

1. Sarah Linn, "The Dunites, Building a Utopia in the Oceano Dunes," *Art Bound*, KCET, July 7, 2013.
2. Jennifer Sharp, "Take Two," KPCC, March 10, 2014.
3. Ann Parker, "Santa Cruz Stories: Ella Thorpe Ellis, Writer, Artist, Dune Child," *Santa Cruz Sentinel*, March 24, 2013.
4. Goddard Jones, "Defender of California's Nipomo Dunes," 19.
5. Schrepfer, "The Nuclear Crucible," 221.
6. Wellock, *Critical Masses*, 73; Wills, *Conservation Fallout*, 2.
7. Wills, *Conservation Fallout*, 38.
8. Wellock, *Critical Masses*, 115.
9. *The Birds* is a 1963 film starring Tippi Hendren as a woman whose purchase of two lovebirds sets off an attack by flocks of birds on residents of Bodega Bay.
10. Charles Hillinger, "Belle's Battle: 'Mother of Ecology' Thrives on Fight with Giant Utility," *Los Angeles Times*, December 17, 1971, A3.
11. Turner, *David Brower*, 109.
12. Wellock, *Critical Masses*, 38–51.
13. Cornell, *Defender of the Dunes*, 41.
14. Goddard Jones, "Defender of California's Nipomo Dunes," 2.

15. Letters to parents, June–September 1926, Goddard Jones Correspondence, folders 4–7.
16. Alma Kitchell, *Let's Talk It Over*, 1938, in Goddard Jones Papers, box 1, folder 4.
17. Goddard Jones, "Defender of California's Nipomo Dunes," 4.
18. William M. Blair, "Echo Park Plan for Dam Dropped," *New York Times*, November 5, 1955, 9; John B. Oakes, "Conservation: A Growing Force," *New York Times*, November 13, 1955, X33. To save Echo Canyon, conservationists agreed not to protest a proposed dam at Glen Canyon, a decision Sierra Club's executive director David Brower later called his biggest regret.
19. Letters to Sierra Club members, 1954, Goddard Jones Papers, box 3, folder 1.
20. Goddard Jones, "Defender of California's Nipomo Dunes," 10–12.
21. Letters from admirers, Goddard Jones Papers, box 9 folder 3.
22. Cornell, *Defender of the Dunes*, 9–11.
23. Goddard Jones Papers, box 1, folder 11.
24. Cornell, *Defender of the Dunes*, 7–14.
25. Cornell, *Defender of the Dunes*, 26.
26. Philip Fradkin, "Which Way for Dunes: Industry Discovers Stretch of State's 'Forgotten' Coast," *Los Angeles Times*, June 15, 1970, C4.
27. Cornell, *Defender of the Dunes*, 29–43.
28. Schrepfer, "The Nuclear Crucible," 220; Cornell, *Defender of the Dunes*, 41–43.
29. Cornell, *Defender of the Dunes*, 53–62; Goddard Jones, "Defender of California's Nipomo Dunes," 21–23.
30. Cornell, *Defender of the Dunes*, 64–68.
31. Glenna Matthews discusses Varian Associates in her book *Silicon Valley*, 127–30.
32. Cornell, *Defender of the Dunes*, 62–63.
33. "Biography of Lee Wilson," Wilson Papers; Goddard Jones Papers, box 1, folder 10.
34. Cornell, *Defender of the Dunes*, 72.
35. Author interview with John Ashbaugh, May 2, 2018, San Luis Obispo.
36. Cornell, *Defender of the Dunes*, 69.
37. Cornell, *Defender of the Dunes*, 69.
38. Cornell, *Defender of the Dunes*, 79–80.
39. Goddard Jones, "Defender of California's Nipomo Dunes," 24. Schrepfer also discusses Fred Eissler's view on nuclear power in "The Nuclear Crucible," 216.
40. Goddard Jones, "Defender of California's Nipomo Dunes," 24.
41. Schrepfer, "The Nuclear Crucible," 214.
42. "They Walked, and Walked, and Walked . . . Then Talked," *San Luis Obispo County Telegram-Tribune*, January 16, 1965, B1.
43. Schrepfer, "The Nuclear Crucible," 215.
44. Schrepfer, "The Nuclear Crucible," 215.

45. Cornell, *Defender of the Dunes*, 108.
46. Goddard Jones Papers, box 3, folder 1, March 1965.
47. Schrepfer, "The Nuclear Crucible," 217.
48. Wellock, *Critical Masses*, 75–78. Goddard Jones Papers, box 2, folder 4 focuses entirely on the battle over Diablo Canyon.
49. Schrepfer, "The Nuclear Crucible," 217.
50. Schrepfer, "The Nuclear Crucible," 217, 218.
51. Goddard Jones Papers, box 4, folder 8.
52. Sierra Club Oral Histories, Kathleen Goddard Jones, 28–30; Cornell, *Defender of the Dunes*, 128.
53. Schrepfer, "The Nuclear Crucible," 222–23.
54. Schrepfer, "The Nuclear Crucible," 229–30.
55. Daryl Lembke, "Hearing Opens Today on PG&E Nuclear Plant," *Los Angeles Times*, February 16, 1967, A15.
56. Gladwin Hill, "Conservatives Win Sierra Club Vote; Director May Be Ousted," *New York Times*, April 17, 1969, 94.
57. *Defender of the Dunes*, 130.
58. Gladwin Hill, "Sierra Club Sending Out Ballots for a Vote Vital to Its Future," *New York Times*, March 14, 1969, 20.
59. "Appeals Filed against Coastal Atomic Facility," *Los Angeles Times*, November 24, 1967, B8.
60. Turner, *David Brower*, 135–46.
61. Turner, *David Brower*, 144.
62. Turner, *David Brower*, 144.
63. Goddard Jones, "Defender of California's Nipomo Dunes," 48.
64. Goddard Jones Papers, box 13, folder 4.
65. Schrepfer, "The Nuclear Crucible," 231–32.
66. Goddard Jones Papers, box 4, folder 14 discusses the Sierra Club moving beyond wilderness.
67. Schrepfer, "The Nuclear Crucible," 230.
68. Cornell, *Defender of the Dunes*, 92, 116–17, 123; Sierra Club Oral Histories, 31–33; "Supporters See Hope for Nipomo Dunes Park," *Los Angeles Times*, January 19, 1966, 25.
69. "Supporters See Hope for Nipomo Dunes Park."
70. Letter to Sierra Club newsletters, including *Peak and Prairie*, Rocky Mountain chapter newsletter, March 1969, Goddard Jones Papers, box 4, folder 13.
71. Cornell, *Defender of the Dunes*, 134.
72. Cornell, *Defender of the Dunes*, 138.
73. Cornell, *Defender of the Dunes*, 154.

74. Shirley Contreras, "Remembering My Friend Kathleen Goddard Jones," *Santa Maria Times*, February 9, 2014.

75. Author interview with John Ashbaugh.

76. Goddard Jones Papers, box 4, folder 16.

77. Cornell, *Defender of the Dunes*, 136–54.

78. Sierra Club Oral Histories, 17.

79. Goddard Jones Correspondence, box 7, folder 9, and box 4, folder 20.

80. Sierra Club Oral Histories, introduction to Kathleen Goddard Jones entry by John Ashbaugh, i. Ashbaugh, now in his seventies, remains active in environmental causes. He teaches history at a local community college and is a former San Luis Obispo City Council member.

81. Sierra Club Oral Histories, introduction to Kathleen Goddard Jones entry by Dirk Walters, ii.

82. Goddard Jones, "Defender of California's Nipomo Dunes," 30.

83. Goddard Jones Papers, box 1, folder 11.

84. Connie Koenenn, "Paradise Preserved: Strange Bedfellows Have United to Protect Guadalupe Nipomo Dunes," *Los Angeles Times*, June 4, 1992, E1. Those signing off on the preservation agreement included the California Parks and Recreation Department, the California Coastal Conservancy, the federal Bureau of Land Management, Land Conservancy of San Luis Obispo County, Santa Barbara County, farmers, oil companies, and Vandenberg Air Force Base.

85. Goddard Jones Papers, box 1, folder 1.

86. Koenenn, "Paradise Preserved."

4. SAVING THE SANTA MONICA MOUNTAINS

1. Weaver, *As I Live and Breathe*, 170. *A Holiday Affair* starred Robert Mitchum and Janet Leigh as a department store clerk and a single mother who fall in love. It was produced by John Hartman for RKO Studios.

2. Weaver, *As I Live and Breathe*, 36, 172, 202–29.

3. Jean Burden, "Where Angels Fear to Tread," *Los Angeles Times*, December 20, 1953, 110.

4. McWilliams, *Southern California*, 234.

5. Art Seidenbaum, "Saving the Mountains," *Los Angeles Times*, June 28, 1972, B1.

6. Robert A. Jones, "Santa Monicas: U.S. Park? Things Are Looking Up," *Los Angeles Times*, October 6, 1977, B1.

7. Lillard, *Eden in Jeopardy*, 279.

8. Matthew W. Roth, "Mulholland Highway and the Engineering Culture of Los Angeles in the 1920s," in Sitton and Deverell, *Metropolis in the Making*, 54–56. The road was named for William Mulholland, a civil engineer largely responsible for building the aqueduct that brought water from Owens Valley to

Southern California. In 1928 Mulholland gained a different kind of notoriety, when St. Francis Dam, built by him near the present-day city of Santa Clarita, suffered catastrophic failure, sending more than twelve billion gallons of water—at times moving fifteen miles an hour—racing down hills and through valleys. More than four hundred people died.

9. Lillard, *Eden in Jeopardy*, 111.
10. John Weaver, "The Fighting Montagnards," *Los Angeles Times*, June 7, 1968, A28.
11. Whittemore, "Zoning Los Angeles."
12. Lillard, *Eden in Jeopardy*, 111–12; "Mopping Up Started in Storm Areas," *Los Angeles Times*, January 20, 1952, A1.
13. Sellers, *Crabgrass Crucible*, 199.
14. Lillard, *Eden in Jeopardy*, 112.
15. Lillard, *Eden in Jeopardy*, 280.
16. "How to Get Rid of Subdividers," Weaver Papers, box 40. Stevens discusses Weaver's involvement in mountain politics in "Feminizing the Urban West."
17. "How to Get Rid of Subdividers," Weaver Papers, box 40.
18. Art Seidenbaum, "The Santa Monica Mountain Hassle: Trouble on Olympus," *Los Angeles Times*, September 20, 1964, 13; Jennifer Audrey Stevens, "Feminizing the Urban West," 134.
19. Lillard, *Eden in Jeopardy*, 109.
20. Weaver Papers, box 40.
21. Stevens, "Feminizing the Urban West," 118.
22. John Weaver, "The Fighting Montagnards."
23. Weaver Papers, box 40.
24. Ursula Vils, "The Battle to Preserve the Mountains," *Los Angeles Times*, November 26, 1978, L1.
25. Nelson Papers, box 1, folder 1, and box 81, folder 2.
26. Nelson Papers, box 1, folder 1; Sarah Dixon, quoted in Nelson obituary, *Los Angeles Times*, May 22, 2003.
27. Nelson Papers, box 17, folder 2.
28. Vils, "The Battle to Preserve the Mountains."
29. Suzanne Guldimann, "Three Women Fought to Create the Santa Monica Mountains Parkland Now Burned by the Woolsey Fire," *Los Angeles Times*, December 2, 2018, B1.
30. Letter from Ellie Oswald to Friends of the Santa Monica Mountains, Nelson Papers, box 17, folder 2.
31. Opinion essay on park, KNXT television, December 1963, Nelson Papers, box 1, folder 8.

32. "Santa Monica Mountains Master Plan Review Offered by Officials," *Los Angeles Times*, July 7, 1963, B20; Al Thrasher, "Master Plan Designed to Ease Conflict," *Los Angeles Times*, June 14, 1964, 71.

33. Doug Smith, "20-Year Battle Pits Conservationists, Developers," *Los Angeles Times*, February 5, 1978, A1.

34. Undated letter, Nelson Papers, box 1, folder 8.

35. Stevens, "Feminizing the Urban West," 149.

36. Lynn Lillston, "Bantamweight Blond Fight to Keep Canyons Uncrowded," *Los Angeles Times*, April 30, 1969, H1.

37. Jeffrey Hansen, "Homeowner Groups: City Hall Fighters Win Voice There," *Los Angeles Times*, June 9, 1974, A1.

38. Letter from Susan Nelson to Robert Jesperson, August 1969, Nelson Papers, box 1, folder 8.

39. Recruitment letter for Friends of the Santa Monica Mountains, Nelson Papers, box 4, folder 1.

40. Summary of city actions related to the Santa Monica Mountains, 1962–1980, Nelson Papers, box 17, folder 2.

41. Harriett Weaver, "The Fearsome Spectre of Brush Fires," *Los Angeles Times*, September 12, 1971, R9.

42. Lillian Melograno, letter to the editor, *Los Angeles Times*, July 24, 1968, B5.

43. Undated letter, Nelson Papers, box 1, folder 10.

44. Valerie J. Nelson, "Margot Feuer Dies at 89; Helped Create Santa Monica Mountains Park," *Los Angeles Times*, June 29, 2012, B4.

45. Vils, "The Battle to Preserve the Mountains"; "Park Service Honors Trailblazing Woman," *Los Angeles Times*, March 28, 1997, A14.

46. Hansen, "Homeowner Groups."

47. Pitt, *A Touch of Wilderness*, 41.

48. Vils, "The Battle to Preserve the Mountains."

49. U.S. Congress, House Committee on Interior and Insular Affairs, Hearings, vol. 1, U.S. Department of Interior, Recreation Advisory Council, "Policy on the Establishment and Administration of Recreation Areas," March 1963.

50. Pitt, *A Touch of Wilderness*, 14.

51. Pitt, *A Touch of Wilderness*, 163.

52. "Santa Monica Mountains Next Target for Zoning Evasion," *Canyon Crier*, August 11, 1969, cited in Stevens, "Feminizing the Urban West," 164.

53. Stevens, "Feminizing the Urban West," 166.

54. Undated letter from Marjorie Braude, Nelson Papers, box 1, folder 8.

55. Pitt, *A Touch of Wilderness*, 114.

56. Dennis McLellan, "Susan Nelson, 76; Mountain Parklands Advocate," *Los Angeles Times*, May 22, 2003, B17.

57. Pitt, *A Touch of Wilderness*, 41.

58. Nelson letter to unnamed recipient, June 1972, Nelson Papers, box 1, folder 10.

59. Testimony before the Senate Subcommittee on Parks and Recreation of the Committee on Interior and Insular Affairs of the U.S. Senate, Ninety-Third Congress, Second session, June 15, 1974, 107, 100, 82.

60. Skip Ferderber, "Hearings Planned on Mountains Park Bill," *Los Angeles Times*, April 18, 1974, SF12.

61. Susan Nelson and Ellen Strote, untitled, undated news article, Nelson Papers, box 4, folder 1.

62. Undated letter, Nelson Papers, box 73, folder 6.

63. Nelson Papers, box 6, folder 25. Proposition 1, the Recreational Lands Bond Act, appeared on the June 1974 California ballot.

64. Additional land came from nearby property owned by Ronald and Nancy Reagan in the 1950s and early 1960s. It is mostly used for horseback riding.

65. Ellen Stern Harris, "Our Shrinking Recreation Areas," *Los Angeles Times*, June 19, 1977, E4.

66. Robert A. Jones, "Santa Monicas, U.S. Park? Things Are Looking Up," *Los Angeles Times*, October 6, 1977, B1.

67. Mary Ellen Strote and Susan Nelson, "Santa Monica Mountains Are Not Out of the Woods," *Los Angeles Times*, July 3, 1977, 11.

68. Pitt, *A Touch of Wilderness*, 127.

69. Jacobs, *A Rage for Justice*, 354.

70. Pitt, *A Touch of Wilderness*, 127.

71. Jacobs, *A Rage for Justice*, 352–57.

72. Pitt, *A Touch of Wilderness*, 120.

73. Vils, "The Battle to Preserve the Mountains."

74. Pitt, *A Touch of Wilderness*, 25.

75. Pitt, *A Touch of Wilderness*, 41–43.

76. Pitt, *A Touch of Wilderness*, 85, 172.

77. Pitt, *A Touch of Wilderness*, 82.

78. Joe Edmiston, executive director of the Santa Monica Mountains Conservancy, quoted in Pitt, *A Touch of Wilderness*, 166.

79. Santa Monica Mountains Comprehensive Plan, February 1979, 3, SMMC. Ca.gov/SMM%20Comprehensive%20Plan.pdf.

80. Santa Monica Mountains Conservancy website.

81. Richard Simon, "Santa Monica Mountain Park Funds Cut Off," *Los Angeles Times*, February 22, 1981, A1. Ironically, in 1966 Reagan had donated a ranch he owned to the Paramount Ranch, which had become part of the SMMNRA.

82. Arthur Eck, quoted in Pitt, *A Touch of Wilderness*, 144.

83. Phillip Gollner, "She Led Fire Safety Campaign: Friends Remember Hillside Crusader," *Los Angeles Times*, March 17, 1989, A12.

84. Elaine Woo, "Hiker Pushed for an L.A. National Park," *Los Angeles Times*, May 23, 2008, B6.

85. Nelson, "Margot Feuer Dies at 89."

86. Pitt, *A Touch of Wilderness*.

87. McLellan, "Susan Nelson, 76."

88. Guldimann, "Three Women Fought."

5. ENVIRONMENTAL JUSTICE

1. Lee, "Community and Union Organizing," 16.

2. Lee, "Community and Union Organizing," 25.

3. Lee, "Community and Union Organizing," 32. The Tenderloin spans an area bounded by Geary Street on the north, Market Street on the south, and Van Ness on the west. It includes Union Square and the downtown Civic Center. It has been somewhat renovated, but still contains many gritty neighborhoods where homeless congregate.

4. Lee, "Community and Union Organizing," 44..

5. The quote is from Winona LaDuke, a member of the Ojibwe tribe of Minnesota and an activist on behalf of Native American and environmental causes. It appears in Stein, *New Perspectives on Environmental Justice*, xiii.

6. Brodkin, *Power Politics*, 52.

7. Melosi, "Environmental Justice," 63–64.

8. "About Clamshell," Clamshell Alliance, accessed June 26, 2020, https:/www.clamshellalliance.net/about.

9. Wills, *Conservation Fallout*, 91, 96–103. Pawel, *The Browns of California*, is the most recent of more than a dozen books on Edmund G. "Pat" Brown Sr. and his son Edmund G. "Jerry" Brown Jr. Another is Rarick's *California Rising*.

10. Evered, *Protest Diablo*, 65–90, 105.

11. Evered, *Protest Diablo*, 98–122. Silkwood's 1974 death has been deemed suspicious, amid rumors that her car was purposely hit from behind and driven off the road. Her family filed a lawsuit against her employer, Kerr-McGee, and won a nearly $2 million settlement. Meryl Streep played Silkwood in the 1983 film of the same name.

12. The name Sojourner Truth was itself a pseudonym, adopted by Isabella Baum-free, who was born a slave in late eighteenth-century New York. After escaping her master, she was taken in by an abolitionist family who purchased her freedom. She became an itinerant preacher, then an abolitionist and women's rights activist.

13. Evered, *Protest Diablo*, 126.

14. Evered, *Protest Diablo*, 47.

15. Pardo, "Mexican American Women Grassroots Community Activists."

16. Gottlieb, *Forcing the Spring*, 266.

17. Dan Morain, "Police Batons Blamed as UFW Official Is Badly Hurt during Bush S.F. Protest," *Los Angeles Times*, September 16, 1988, B3. The removal of mostly Mexican American families from Chavez Ravine provides another example of differential treatment. While white residents of the Santa Monica Mountains in Los Angeles worked with city and county officials on plans to save the moun-tains from overdevelopment, residents of nearby Chavez Ravine were forcibly evicted from their homes to make way for the newly arrived Los Angeles Dodgers. Few officials, or mountain residents, showed any sympathy for those displaced. Dodger Stadium opened in 1962.

18. In a class action lawsuit, plaintiffs in *Bean v. Southwestern Waste* argued that the company's decision to place a solid waste facility in a minority Houston neighborhood represented a violation of the U.S. Constitution's Fourteenth Amendment equal protection clause because it was racially based. They sought a preliminary injunction and a temporary restraining order to stop the company from proceeding. The request was denied. According to the ruling, more than half the permits issued for such facilities came in nonminority neighborhoods.

19. Robert R. M. Verchick discusses Hazel Johnson's role in the movement in "Feminist Theory and Environmental Justice," in Stein, *New Perspectives on Envi-ronmental Justice*, 78–81.

20. Cynthia Warrick, "History of the Environmental Justice Move-ment," ResearchGate, October 2015, 1, https://www.researchgate.net/ publication/282660631_History_of_the_Environmental_Justice_Movement.

21. Brodkin, *Power Politics*, 52.

22. Nancy C. Unger, "Women, Sexuality, and Environmental Justice in American History," in Stein, *New Perspectives on Environmental Justice*, 57. On the Pine Ridge Reservation in South Dakota, for example, Native women suffered miscarriages at six times the overall national rate.

23. Lee, "Community and Union Organizing," 49.

24. Gottlieb, *Forcing the Spring*, 257–58.

25. Gottlieb, *Forcing the Spring*, 3.
26. Melosi, "Environmental Justice," 64.
27. Verchick, "Feminist Theory and Environmental Justice," 65.
28. Gottlieb, *Forcing the Spring*, 306.
29. Shah, *Laotian Daughters*, 4.
30. Gregory Dicum, "GREEN/Marie Harrison and the Fight for Bayview-Hunters Point," *SFGATE*, February 2, 2005, https://www.sfgate.com/homeandgarden/article/GREEN-Marie-Harrison-And-The-Fight-For-2733707.php.
31. Dicum, "GREEN/Marie Harrison and the Fight for Bayview-Hunters Point."
32. Thomas Nahmyo, "More Americans Grow Old in Toxic Environments," *Washington Informer*, June 3, 2010, 1, 5.
33. Dicum, "GREEN/Marie Harrison and the Fight for Bayview-Hunters Point." Greenaction's reach extends beyond the Bay Area. The group also is working with activists in Kettleman City, along the I-5 corridor in California's Central Valley. The mostly Latino/a population there suffers from contaminated drinking water from oil fields and dumped sewage from nearby farms.
34. Tony Kelly and Marie Harrison, "Trying to Build a Future on Toxic Ground," *San Francisco Chronicle*, February 13, 2018.
35. Brodkin, *Power Politics*, 13–14.
36. Builders of one freeway, the Century (105)—which runs through South Gate and other lower income communities—razed entire neighborhoods in the late 1960s and early 1970s. The area remained vacant for a decade before construction began.
37. Pardo, "Mexican American Women Grassroots Community Activists," 2.
38. The Mexican American Political Association was formed in 1960 to promote the election of Latinos/as at the local and state level. The Chicano movement of the 1970s aided in this effort, as did the emergence of the Latino Caucus and the Black Caucus, but it was not until the 1980s that minorities began to see electoral success. At about the same time, the increasing power of Assembly Speaker Willie Brown fueled a movement to set term limits for statewide lawmakers. In 1990, after Brown had served more than a decade as speaker, voters limited assembly members to three two-year terms and state senators to two four-year terms. Brown went on to become mayor of San Francisco. The Asian Pacific Islander caucus was formed in 2001.
39. Pardo, "Mexican American Women Grassroots Community Activists," 3.
40. Pardo, "Mexican American Women Grassroots Community Activists," 4.
41. Kevin Roderick, "Vernon Incinerator Is Shouted Down—for a Month, at Least," *Los Angeles Times*, December 5, 1987, A1.
42. "Mothers Group Fights Back in Los Angeles," *New York Times*, December 5, 1989, A32.

43. Maura Dolan, "Toxic Waste Incinerator Bid Abandoned," *Los Angeles Times*, May 24, 1991, A1.

44. Pardo, "Mexican American Women Grassroots Community Activists," 6.

45. Richard Marosi, "The Mountain Is Crumbling at Long Last," *Los Angeles Times*, May 1, 2001, B1.

46. Brodkin, *Power Politics*, 53.

47. Simon Romero, "Concrete Recycler Wins Renewal Vote," *Los Angeles Times*, November 20, 1994, B10.

48. Esperanza Marquez died in 2017 at the age of ninety-three. Before her death she received many honors from her community and from state lawmakers. She also was the subject of study at UCLA and the focus of a French documentary.

49. Saika, "Voices of Feminism Oral History Project," 4–30.

50. Saika, "Voices of Feminism Oral History Project," 34.

51. Shah, *Laotian Daughters*, 1–21.

52. Paul Feldman, "Toxic Runoff Made Teacher an Activist," *Los Angeles Times*, July 31, 1992, A24; Gottlieb, *Forcing the Spring*, 163–64.

53. Gottlieb, *Forcing the Spring*, 164.

54. Gottlieb, *Forcing the Spring*, 207.

55. Feldman, "Toxic Runoff Made Teacher an Activist."

56. Tom Gorman, "A Tainted Legacy: Toxic Dump Site in Riverside County Has Sparked the Nation's Largest Civil Suit," *Los Angeles Times*, January 10, 1993, p. A3.

57. Tom Gorman, "Final Settlement Is Approved in Waste Dump Case," *Los Angeles Times*, November 18, 1994, A3.

58. "Our Roots," Center for Community Action and Environmental Justice, accessed June 26, 2020, https://www.ccaej.org/our-roots.

59. Stein, introduction, in *New Perspectives on Environmental Justice*, 4, 11.

60. Bullard, "Environmental Justice in the 21st Century," 151.

61. The film *Erin Brockovich* was released by Universal Studios in 2000. It starred Julia Roberts in the title role and Albert Finney as her boss, Ed Masry, partner in the law firm that represented plaintiffs in the class action lawsuit. It was directed by Steven Soderbergh.

62. Albert Huang, "The 20th Anniversary of President Clinton's Executive Order 12898 on Environmental Justice," NRDC, February 10, 2014, https://drrobertbull-ard.com/new-report-tracks-environmental-justice-movement-over-five-decades/.

63. State Senator Connie M. Leyva represents California's 20th Senatorial District, which includes the cities of Chino, Colton, Fontana, Ontario, and Pomona.

64. "Summary of Executive Order 12898—Federal Actions to Address Environmental Justice in Minority Populations and Low-Income Populations," 59 FR7629, February 16, 1994, in "Presidential Documents," *Federal Register* 59, no. 32

(February 16): 1994. The November 2019 presidential forum on environmental justice was hosted by Amy Goodman of Democracy Now and sponsored by the National Black Caucus of state legislatures, civil rights, youth, and environmental groups.

65. Huang, "The 20th Anniversary of President Clinton's Executive Order 12898 on Environmental Justice."

66. Steve Lopez, "Activist Takes on Prop. 23; Woman Fights Big Oil over Delaying the State's Global Warming Law," *Los Angeles Times*, September 26, 2010, A2.

CONCLUSION

1. Center for American Progress, "Environmental Justice and National Environmental Groups Advance a Historic Joint Climate Platform," July 18, 2019, www.americanprogress.org.

2. Carson, *Silent Spring*, 277.

BIBLIOGRAPHY

ARCHIVES AND MANUSCRIPT MATERIALS

Bade, Elizabeth Marston. "Recollections of William F. Bade and the Early Sierra Club." Interview by Eleanor Bade, April 1976. Oral History Center, Bancroft Library, University of California, Berkeley.

Bodovitz, Joseph. "Joe Bodovitz: Founding Director of the Bay Conservation Development Commission and the California Coastal Commission." Interview by Martin Meeker, 2015. Oral History Center, Bancroft Library, University of California, Berkeley.

Evans, Nora. "Sixty Years with the Sierra Club." Interview by Judy Snyder, October 1972. Oral History Center, Bancroft Library, University of California, Berkeley.

Farquhar, Marjory. "Pioneer Woman Rock Climber and Sierra Club Director." Interview by Ann Lage, April 1977. Oral History Center, Bancroft Library, University of California, Berkeley.

Goddard Jones, Kathleen. Correspondence (MS 173). Special Collections and Archives, California Polytechnic State University, San Luis Obispo CA.

———. "Defender of California's Nipomo Dunes, Steadfast Sierra Club Volunteer." Interview by Anne Van Tyne, 1984. Sierra Club Nationwide II, Regional Oral History Office, University of California, Berkeley.

———. Papers (MS 119). San Luis Obispo County Environmental Archives, California Polytechnic State University, San Luis Obispo CA.

Green, Dorothy. "Environmental Activism in Los Angeles." Interviews by Jane Collings, March–July 2006. Center for Oral History Research, University of California, Los Angeles.

Gulick, Esther, Catherine Kerr, and Sylvia McLaughlin. "Save the San Francisco Bay Association, 1961–1986." Interview by Malca Chall, 1985, 1986. Regional Oral History Office, Bancroft Library, University of California, Berkeley.

Kittredge, Janice Rivers. "Volunteers and Employment Careers: Save the San Francisco Bay Association." Interview by Malca Chall, 1998–99. Regional Oral History Office, Bancroft Library, University of California, Berkeley.

Lee, Pamela Tau. "Community and Union Organizing, and Environmental Justice in the San Francisco Bay Area, 1967–2000." Interview by Carl Wilmsen, 2003. Regional Oral History Office, Bancroft Library, University of California, Berkeley.

McLaughlin, Sylvia. "Citizen Activist for the Environment: Saving San Francisco Bay, Promoting Shoreline Parks and Natural Values in Urban and Campus Planning." Interviews by Ann Lage, 2006–7. University of California, Berkeley, Bancroft Library.

Mothers for Peace. Records of the San Luis Obispo Mothers for Peace (MS0195). Special Collections and Archives, California Polytechnic State University, San Luis Obispo CA.

Nelson, Susan B. Papers (URB/SBN). Special Collections and Archives, California State University, Northridge.

Petris, Nicholas. "Oral History Interview with Nicholas Petris, California State Senator, 1967–, California State Assemblyman, 1959–1966." Interview by Gabrielle Morris, October 12, 1989. University of California, Berkeley, Bancroft Library.

Pitt, Leonard. *A Touch of Wilderness: Oral Histories on the Formation of the Santa Monica Mountains National Recreation Area*. Los Angeles CA: National Park Service, 2015.

Saika, Peggy. "Voices of Feminism Oral History Project, Sophia Smith Collection." Interview by Loretta Ross, February 20, 2006. Smith College, Northampton MA.

Sierra Club Oral Histories. "Sierra Clubwomen, I, II, and III," 1976–77, 1983. Regional Oral History Series, Bancroft Library, University of California, Berkeley.

Weaver, Harriet. Papers (MS 1246). UCLA Special Collections, Charles E. Young Research Library, University of California, Los Angeles.

Wilson, Lee. Papers (MS 113). Special Collections, Cal Poly, San Luis Obispo.

PUBLISHED WORKS

Alexiou, Alice Sparberg. *Jane Jacobs, Urban Visionary*. New York: Harper Perennial, 2010.

Austin, Mary. *Land of Little Rain*. New York: Houghton Mifflin Harcourt, 1903.

Baranzini, Marlene Smith. *The Shirley Letters: From the California Gold Mines, 1851–52.* San Francisco: Heyday Press, 2014.

Blum, Howard. *American Lightning: Terror, Mystery, the Birth of Hollywood, and the Crime of the Century.* New York: Broadway Books, 2008.

Bonta, Marcia Myers. *American Women Afield: Writings by Pioneering Women Naturalists.* College Station: Texas A&M University Press, 1995.

Brodkin, Karen. *Power Politics: Environmental Activism in South Los Angeles.* New Brunswick NJ: Rutgers University Press, 2009.

Bryson, Bill. *The Life and Times of the Thunderbolt Kid.* New York: Random House, 2006.

Bullard, Robert. "Environmental Justice in the 21st Century: Race Still Matters." *Phylon* 3–4 (Winter 2001): 151–71.

Carson, Rachel. *Silent Spring.* 40th anniversary edition. New York: Houghton Mifflin, 2002.

Cherny, Robert W., Mary Ann Irwin, and Ann W. Wilson, eds. *California Women and Politics: From the Gold Rush to the Great Depression.* Lincoln: University of Nebraska Press, 2011.

Cole, Luke W. *From the Ground Up: Environmental Racism and the Rise of the Environmental Justice Movement.* New York: NYU Press, 2001.

Cornell, Virginia. *Defender of the Dunes: The Kathleen Goddard Jones Story.* Carpenteria CA: Manifest, 2001.

Davis, Mike. *City of Quartz: Excavating the Future in Los Angeles.* New York: Vintage, 1990.

Didion, Joan. *Play It as It Lays.* Kindle edition, 2017.

———. *Slouching toward Bethlehem.* New York: Farrar, Strauss and Giroux, 1968.

Dolezel, Janine, and Bruce N. Warren. "A Case Study in Environmental Legislation." *Stanford Law Review,* January 1971, 349–66.

Douglas, Marjory Stoneman. *The Everglades: River of Grass.* New York: Rinehart, 1947.

Douglas, Marjory, and John Rothchild. *Voice of the River.* Kindle edition, 1987.

Epstein, Barbara. *Political Protest and Cultural Revolution: Nonviolent Direct Action in the 1970s and 1980s.* Berkeley: University of California Press, 1993.

Evered, Judith. *Protest Diablo: Living and Dying under the Shadow of a Nuclear Power Plant.* Scotts Valley CA: CreateSpace, 2010.

Gabler, Neal. *Walt Disney: The Triumph of the American Imagination.* New York: Random House, 2006.

Goodman, Susan, and Carl Dawson. *Mary Austin and the American West.* Berkeley: University of California Press, 2009.

Gottlieb, Robert. *Forcing the Spring: The Transformation of the Environmental Movement.* Washington DC: Island Press, 1993.

Gulick, Esther, Catherine Kerr, and Sylvia McLaughlin. "Saving the San Francisco Bay, Past, Present, and Future." Albright Lecture, College of Natural Resources, University of California, Berkeley, August 14, 1988.

Hoffman, Abraham. "Mary Austin, Stafford Austin, and the Owens Valley." *Journal of the Southwest* 53, nos. 3–4 (Autumn–Winter 2011): 305–22.

Hurley, Andrew. *Environmental Inequalities: Class, Race, and Industrial Pollution in Gary, Indiana.* Chapel Hill: University of North Carolina Press, 1995.

Jackson, Kenneth T. *Crabgrass Frontier: The Suburbanization of the United States.* New York: Oxford University Press, 1987.

Jacobs, Chip, and William J. Kelley. *Smogtown: The Lung-Burning History of Pollution in Los Angeles.* New York: Overlook Press, 2009.

Jacobs, John. *A Rage for Justice: The Politics and Passion of Phillip Burton.* Berkeley: University of California Press, 1995.

Johnson, Katherine M. "Captain Blake versus the Highwaymen: Or How San Francisco Won the Freeway Revolt." *Journal of Public History* 8, no. 1 (October 2008): 56–83.

Kaufman, Polly Welts. *National Parks and the Woman's Voice.* Albuquerque: University of New Mexico Press, 2006.

Laslett, John H. M. *Shameful Victory: The Los Angeles Dodgers, the Red Scare, and the Hidden History of Chavez Ravine.* 2nd edition. Tucson: University of Arizona Press, 2015.

Lee, Erika, and Judy Yung. *Angel Island: Immigrant Gateway to America.* New York: Oxford University Press, 2012.

Lillard, Richard. *Eden in Jeopardy, Man's Prodigal Meddling with His Environment: The Southern California Experience.* New York: Alfred A. Knopf, 1966.

Lynn, Susan. *Progressive Women in Conservative Times: Racial Justice, Peace, and Feminism, 1945 to the 1960s.* New Brunswick NJ: Rutgers University Press, 1992.

Matthews, Glenna. *Silicon Valley, Women, and the California Dream: Gender, Class, and Opportunity in the Twentieth Century.* Palo Alto CA: Stanford University Press, 2002.

McClain, Molly. *Ellen Browning Scripps: New Money and American Philanthropy.* Lincoln: University of Nebraska Press, 2018.

McWilliams, Carey. *Southern California: An Island on the Land.* New York: Peregrine Smith, 1980.

Melosi, Martin. "Environmental Justice, Political Agenda Setting, and Myths of History." *Journal of Policy History* 12, no. 1 (2000): 43–71.

Merchant, Carolyn. *Earthcare: Women and the Environment.* New York: Routledge, 1995.

———. *Ecological Revolutions: Nature, Gender, and Science in New England.* 2nd edition. Chapel Hill: University of North Carolina Press, 2010.

———. "Women of the Progressive Conservation Movement: 1900–1916." *Environmental Review* 8, no. 1 (Spring 1984): 57–85.

Meyerowitz, Joanne. "Beyond the Feminine Mystique: A Reassessment of Postwar Mass Culture, 1946–1958." *Journal of American History* 79, no. 4 (March 1998): 1445–82.

Morrow, Greg. "The Homeowner Revolution: Democracy, Land Use and the Los Angeles Slow Growth Movement, 1965–1992." PhD diss., University of California, Los Angeles, 2013.

Murie, Margaret. *Two in the Far North*. Portland OR: Alaska Northwest Books, 2003.

Napoli, Lisa. *Ray and Joan: The Man Who Made the McDonald's Fortune and the Woman Who Gave It All Away*. New York: Dutton, 2016.

Nash, Roderick F. *Wilderness and the American Mind*. 5th edition. New Haven CT: Yale University Press, 2014.

Odell, Rice. *The Saving of San Francisco Bay: A Report on Citizen Action and Regional Planning*. Washington DC: Conservation Foundation, 1972.

Pardo, Mary. "Mexican American Women Grassroots Community Activists: Mothers of East Los Angeles." *Frontiers: A Journal of Women's Studies* 1 (1990): 1–7.

Pawel, Miriam. *The Browns of California: The Family Dynasty That Transformed a State and Shaped a Nation*. New York: Bloomsbury, 2018.

Pincetl, Stephanie. "The Peculiar Legacy of Progressivism: Claire Dedrick's Encounter with Forest Practices Regulation in California." *Forest and Conservation History*, January 1990, 26–34.

Rarick, Ethan. *California Rising: The Life and Times of Pat Brown*. Berkeley: University of California Press, 2005.

Reisner, Marc. *Cadillac Desert: The American West and Its Disappearing Water*. New York: Penguin, 1986.

Riley, Glenda. *Women and Nature: Saving the "Wild" West*. Lincoln: University of Nebraska Press, 1984.

Scharff, Virginia, ed. *Seeing Nature through Gender*. Lawrence: University Press of Kansas, 2003.

Schrepfer, Susan R. *Nature's Altars: Mountains, Gender, and American Environmentalism*. Lawrence: University Press of Kansas, 2005.

———. "The Nuclear Crucible: Diablo Canyon and the Transformation of the Sierra Club, 1965–1985." *California History* 71, no. 2 (1992): 212–37.

Scott, Mel. *The Future of San Francisco Bay*. Berkeley: Institute for Government Studies, University of California–Berkeley, 1963.

Scott, Mel. *The San Francisco Bay Area: A Metropolis in Perspective*. Berkeley: University of California Press, 1985.

Sellers, Christopher. *Crabgrass Crucible: Suburban Nature and the Rise of Environmentalism in the 20th Century.* Chapel Hill: University of North Carolina Press, 2012.

Shah, Bindi V. *Laotian Daughters: Working toward Community, Belonging, and Environmental Justice.* Philadelphia PA: Temple University Press, 2011.

Sitton, Tom, and William Deverell. *Metropolis in the Making: Los Angeles in the 1920s.* Berkeley: University of California Press, 2001.

Smith, Michael L. *Pacific Visions: California Scientists and the Environment, 1850–1915.* New Haven CT: Yale University Press, 1987.

Souder, William. *On a Farther Shore: The Life and Legacy of Rachel Carson.* New York: Crown Books, 2012.

Starr, Kevin. *Americans and the California Dream, 1850–1915.* London: Oxford University Press, 1986.

———. *Golden Dreams: California in the Age of Abundance, 1950–1963.* London: Oxford University Press, 2009.

Stein, Rachel, ed. *New Perspectives on Environmental Justice: Gender, Sexuality, and Activism.* New Brunswick NJ: Rutgers University Press, 2004.

Stevens, Jennifer A. "Feminizing the Urban West: Green Cities, and Open Space in the Post-War Era, 1950–2000." PhD diss., University of California, Davis, 2008.

Swerdlow, Amy. *Women Strike for Peace: Traditional Motherhood and Radical Politics in the 1960s.* Chicago: University of Chicago Press, 1993.

Thiessen, Janet. *Dorothy Erskine: Graceful Crusader for Our Environment.* Sonoma CA: Dorothy Erskine Biography LLC, 2010.

Turner, Tom. *David Brower: The Making of the Environmental Movement.* Berkeley: University of California Press, 2015.

Unger, Nancy C. *Beyond Nature's Housekeepers: American Women in Environmental History.* New York: Oxford University Press, 2012.

Valle, Victor. *City of Industry: Genealogies of Power in Southern California.* New Brunswick NJ: Rutgers University Press, 2009.

Wachs, Martin. "Autos, Transit, and the Sprawl of Los Angeles in the 1920s." *Journal of the American Planning Association* 50, no. 3 (1984): 297–310.

Waldie, D. J. *Holy Land: A Suburban Memoir.* New York: W. W. Norton, 1996.

Walker, Richard A. *The Country in the City: The Greening of the San Francisco Bay Area.* Seattle: University of Washington Press, 2008.

Weaver, John D. *As I Live and Breathe.* New York: Rinehart, 1959.

Wellock, Thomas. *Critical Masses: Opposition to Nuclear Power in California, 1958–78.* Madison: University of Wisconsin Press, 1998.

Whitfield, Eileen. *Pickford: The Woman Who Made Hollywood.* Lexington: University Press of Kentucky, 2007.

Whittemore, Andrew H. "Zoning Los Angeles: A Brief History of Four Regimes."
  Planning Perspectives 27 (June 2012): 393–415.
Wills, John. *Conservation Fallout: Nuclear Protest at Diablo*. Reno: University of
  Nevada Press, 2006.
Worster, Donald. *A Passion for Nature: The Life of John Muir*. London: Oxford Univer-
  sity Press, 2011.

INDEX

Page numbers in italics indicate illustrations.

Abalone Alliance, 103, *139*, 140
activism: about, 2, 3, 4; antinuclear, 139, 140, 141; environmental, 114–15; family support for, 6–7; political, 62, 139. *See also* female activists
Adams, Ansel, 53, 59, 77, 80, 86, 94–96
Adams, Janet, 66, 67
Aggregate Recycling, 150, 151
air pollution, 2, 13–16, 51, 130, 145, 165n2
Air Quality Act, 16
Alexander, Annie, 25
Anderson, Glenn, 54, 58, 59
Angel Island, 32–33
antinuclear activism, 139, 140, 141
Arctic National Wildlife Refuge, 4, 22
Army Corps of Engineers, 48, 56
Ashbaugh, John, 88, 99, 100, 175n80
Asian Pacific Environmental Network (APEN), 151, 152

Atomic Energy Commission, 38, 78
Audubon Society, 32, 50, 71, 82, 99
Austin, Mary Hunter, 23–25

Bade, Elizabeth Marston, 27
Bay Conservation Development Commission (BCDC), 57, 60, 61, 64, 70
bay development, 56–58, 62
Bayview–Hunters Point, 144, 145, 162, 181n33
Berkeley Bay plan, 49–50
Berkeley City Council and Planning Commission, 51, 52, 54
The Birds (Hitchcock), 78
Bodega Bay, 37, 51, 78, 86, 172n9
Brandegee, Mary Katherine, 25
Brockovich, Erin, 156
Brower, David, 81, 82, 93–96
Brown, Edmund G., 15, 39, 61–63, 161

Brown, Willie, 148, 181n38
brushfires, 115, 120
builders: and bay development, 57–58; of freeways, 3, 181n36; and home-building, 17–18, 110–12; removing land from, 114, 117
Burns, Hugh, 67, 68
Burton, Phillip, 127–28

California: fast-food chains in, 19; freeway construction in, 18–19; homebuilding in, 17–18, 110–12; Master Plan for Higher Education for, 49, 52; rise in population of, 16–17; statehood of, 47, 55
California Academy of Sciences, 25, 26
California Coastal Commission, 40, 69, 129, 161
California Federation of Women's Clubs (CFWC), 28, 29
California Native Plant Society, 5, 99, 101
California Public Utilities Commission, 70, 78, 94
Carson, Rachel, 4, 8, 21, 141, 159, 162
Chall, Malca, 70
Chavez Ravine neighborhood, 8, 180n17
Citizens for Regional Recreation and Parks (CRRP), 34, 50, 54
civil rights, 40, 63, 142, 151
Clamshell Alliance, 139
Coastal Conservancy, 101, 175n84
Communities for a Better Environment (CBE), 150, 151
communities of color, 10, 143, 150
community organizations, 4, 5, 146, 149
Conservation Associates, 90, 92, 97
conservation groups, 20, 50, 65, 79, 82
construction: of dams, 2, 20, 22, 35; freeway, 18–19, 38, 147; grading rules

for, 112; nuclear power plant, 89, 96; protests about stopping, 114; residential, 110–13
contaminated areas, 145–46, 154–55

dam building: about, 2, 20, 22; campaign against, 35, 65, 81–82; referendum in favor of, 29
Dearing, Betty Brown, 6, 7, 114, 118, 122
The Death and Life of Great American Cities (Jacobs), 22
Dedrick, Claire, 5, 53, 66–67, 69, 161
Depression, 18, 46
Diablo Canyon, 2, 92–97, 100–103, 139
Diercks, Kenneth, 86–87, 89
Dinosaur National Park, 20, 35, 82
Dolwig, Richard, 67, 69
Douglas, Marjory Stoneman, 4, 21
Dunites, 77, 80, 99, 102

earthquake fault, 38, 79, 96
earthquakes, 9, 26, 34, 38, 79, 149
Eastwood, Alice, 26
Echo Park, 132, 173n18
Edwards, Lucretia, 39–40
Eissler, Fred, 89, 92, 93
environmental activism, 64, 114–15, 144, 158, 169n83
environmental campaigns, 2, 16, 37, 113, 115
environmental groups, 64–67, 162
environmentalism, 3, 4, 20, 35, 65, 138
environmental justice: and Abalone Alliance, 103, 139–40; about, 9, 10, 132, 136, 162; and Clamshell Alliance, 139; conference related to, 143; impact of, 157; and maximum security prison, 140, 148–49; media attention for, 156; and non-Anglo politicians, 147–48; and racism, 141; and Sierra Club, 144,

156; and Stringfellow Acid Pits, 153, 154, 155; victories achieved through, 155; women's role in, 144–45

environmental legislation, 67, 69, 161

environmental movement: about, 3, 4; conclusion about, 159–63; consensus in favor of, 123; and protesters, 63, 112, 139–40; and public relations campaigns, 4, 66, 91, 93, 141, 161; and racism, 141, 143, 155, 162; women's role in, 21, 23–24, 35–36. *See also* activism

environmental organizations, 5, 65, 86, 99, 101

Erskine, Dorothy, 5–6, 33–35, 50, 65

Evans, Nora, 26–27

Evered, Judith, 140

Everglades, 4, 21, 22

Farquhar, Marjory, 30–31

Federation of Hillside and Canyon Associations, 5, 6, 8, 111, 118

female activists: about, 4–5; agendas of, 20; awareness about, 55; and city beautification projects, 36; and coastline access to public, 39–40; conclusion about, 160–62; and creation of parks, 32; elite status of, 8, 53; family support for, 6–7; and Hetch Hetchy damming, 29; influence of, 58, 62; lobbying campaigns by, 2, 38–39, 85, 124; opinion polls about, 37; outreach efforts by, 54–55; and pesticides, 21; and Progressivism, 27–28; and Proposition 20, 40; and renewal projects, 22; and sequoias, 28–29; and smog issue, 15–16; of SOS group, 13–15; traditional appearance of, 7, 52; and urban planning, 34;

views about, 58; volunteering by, 36, 41; and water wars, 24–25

Feminine Mystique (Friedan), 5, 16

Feuer, Margot, 5, 106, 121, 124–25, 132

fires and floods, 113, 120, 130, 152, 160

freeways: about, 2, 3, 181n36; construction of, 18–19, 38, 147; media attention for, 38–39; protests about, 38–39; and Santa Monica Mountains, 121; support for, 39; and traffic issue, 19

Friedan, Betty, 5, 16

Friends of the Santa Monica Mountains, Parks and Seashore, 116, 117, 121–23

The Future of San Francisco Bay (Scott), 55

Gaffney, Rose, 78–79

garden clubs, 15, 54

gender inequities, 21, 137

Gilliam, Harold, 50, 51, 53, 72, 172n87

Goddard, Kathleen. *See* Jones, Kathleen Goddard

Golden Gate Bridge, 31, 47, 56, 168n58

grading rules, 112

Green, Dorothy, 40–42

Greenaction, 146

Greenbelt Alliance, 34, 65, 69

Gulick, Esther: about, 1, 5, 6, 8; and Berkeley Bay plan, 49–50; and Catherine Kerr, 49; death of, 71; interview of, 70–71; meeting attended by, 52–53; photo of, *44*

Gutierrez, Juana, 147, 148, 149

Harrison, Marie, *136*, 144, 146, 181n33

Harwood, Aurelia, *12*, 30

hazardous waste, 138, 139, 141, 145, 149, 154

Heal the Bay organization, 41–42, 169n83

Hearst Castle, 76

Hetch Hetchy damming, 20, 29, 85

hikes and hiking: about, 27, 30; by Jill Swift, 122, 124, 125; by Kathleen Jones, 99–100; at Nipomo Dunes, 84, 87, 99, 100; publicity about, 91; Sierra Club, 76, 84

hillsides, 19, 110, 111, 113, 132, 160

homebuilding, 17–18, 110–12

Huerta, Dolores, 8, 142

Hunters Point Power Plant, 145, 146, 162

Hyperion Sewage Treatment Plant, 41, 169n83

incinerators, 9, 138, 141, 147, 149, 156

industrial contamination, 14, 144, 150, 162

Interstate Highway Act, 18

Jackson, Duncan, 81, 82, 83

Jackson, Kathleen Goddard. *See* Jones, Kathleen Goddard

Johnson, Hazel, 142

Jones, Gaylord, 99, 101

Jones, Kathleen Goddard: about, 1–2, 4–5, 7; and dam building campaign, 82; death of, 102; divorce of, 7, 81, 97; early life of, 80; and Kenneth Diercks, 86–87; and Lee Wilson, 87–88, 90; marriage of, 81, 99; meetings attended by, 88–89; photo of, 74; Sierra Club work, 81–82, 85; tribute to, 102; trip through high Sierras, 98–99; trip to Europe, 80; weekly hikes of, 99–100

Kerr, Catherine "Kay": about, 1, 4, 5, 6, 8; and Berkeley Bay plan, 49–50; death of, 72; and Esther Gulick, 49; interview of, 70–71; photo of, 44; responsibilities of, 51

Kerr, Clark, 45, 48–49, 62–63

Kittredge, Janice, 8, 59

Knox, John T., 66

Kortum, Jean, 37–38

La Causa, 150, 151

landfills, 3, 56, 57, 120, 142, 156

Land of Little Rain (Austin), 24

landowners, 40, 101, 114, 121, 129

land use, 57, 70, 143

League of Women Voters, 6, 15, 39, 54

Lee, Pamela Tau, 137–39, 143

Leyva, Connie, 156

Life and Letters of John Muir (Bade), 27

Litton, Martin, 92, 93

Livermore, Caroline Sealy, 32–33, 68

lobbying campaigns, 2, 38–39, 85, 124

local governments, 62, 68, 78, 125, 157

logging and loggers, 2, 3, 29, 30

Los Angeles: as a center for defense industry, 108–9; massive growth in, 109–10. *See also* California; Santa Monica Mountains

Los Angeles City Council, 7, 119–22, 130, 148, 151

Los Angeles Dodgers, 8, 19, 166n16, 180n17

lower-income residents, 111, 132, 157, 162

low-income neighborhood, 144, 146, 150, 155

Marin Conservation League, 6, 32, 33, 34

Marquez, Esperanza "Linda," 149, 150, 151, 182n48

Master Plan for Higher Education, 49, 52

maximum security prison, 140, 147–49

McAteer, Eugene, 58, 59, 60, 61, 62

McDonald brothers, 166n15

McLaughlin, Don, 45, 46, 53, 62–63, 70

McLaughlin, Sylvia: about, 1, 5; and Berkeley Bay plan, 49–50; death of, 72; dressing style of, 52; interview of, 70–72; marriage of, 47; meetings attended by, 52–53, 68; move to Berkeley, 45–46; outreach efforts by, 54; photo of, *44*; privileged life of, 46–47; responsibilities of, 51; volunteering by, 36

media attention: for environmental justice, 156; for freeway proposals, 38–39; for Heal the Bay organization, 41; for San Francisco Bay, 60–61; for women of SOS, 14

Melograno, Lillian, 113, 118, 121

Mothers of East L.A., 9, 147, 149

mountain activists, 116, 117, 124

mountain residents, *106*, 111, 112, 113, 116, 180n17

Muir, John, 2, 20, 27, 85

Mulholland, William, 175n8, 176n8

Mulholland Drive, 110, 120, 121, 122

multifamily dwellings, 8, 111

Murie, Margaret "Mardy," 4, 22

National Park Service, 30, 84, 124, 128, 131

National People of Color Environmental Leadership Summit, 143, 152

natural resources, 3, 27–28, 55, 60, 101, 114

Nature Conservancy, 32, 101, 102

Nelson, Sue, 5, *106*, 115–16, 124–26, 128–30, 132–33

Newman, Penny, 153–55

Nipomo Dunes: about, 1, 4, 10; alliance for preserving, 101–2; designation of, 100; and Dunites, 77, 80, 99, 102; hiking at, 84; history of, 76–77; nuclear power plant on, 75–76, 84–85, 91–92; proposed park for, 97. *See also* Sierra Club

non-Anglo politicians, 147–48

nuclear power, 20, 35, 78, 96

nuclear power plant(s): about, 1, 3, 7; campaign against building, 37–38, 74, 79, 84–85; and Clamshell Alliance, 139; cleanup issues, 146; construction of, 89, 96; and Diablo Canyon, 92–93; on Nipomo Dunes, 75–76, 84–85, 91–92; shutdown of, 103, 146

nuclear testing, 20, 23

off-road vehicles, 88, 100, 102

oil drilling, 35, 131

open spaces, 8, 19, 31, 33–35, 124

Owens Valley water wars, 24–25

Pacific Gas and Electric Company, 1, 37–38, 75, 77–79, 90

parks: boundary issue, 129–30; legislation related to, 127–28; planning for, 115–16, 130; preserving land for, *106*, 116–17, 123, 128–31; state, 29, 116, 119, 126, 128–29

People for Open Space, 5, 34, 71

people of color, 138, 141, 143, 144, 156

pesticides, 4, 8, 21, 35, 70, 141–42

Petris, Nicholas, 57, 58, 60, 61, 62

Play It as It Lays (Didion), 18

political activism, 62, 139

pollution: air, 2, 13–16, 51, 130, 145, 165n2; emissions of, 124; toxic, 146, 152, 153, 157; water, 35, 69, 159

property owners, 118, 121, 131

Proposition 1, 97

Proposition 20, 40, 69

Proposition 23, 157, 158

Proposition 227, 152–53

protesters, 63, 112, 139–40

public relations, 4, 66, 91, 93, 141, 161

racism, 141, 143, 155, 162

raw sewage, 40, 48, 71

Reagan, Ronald, 16, 62–63, 68–69, 178n64, 179n81

recreational use, 91, 92, 117

recreation area, 114, 123, 126, 128–31, 133, 161

redwoods, 3, 29, 30

residential construction, 110–13

River of Grass (Douglas), 21

Rock, Julie Rynerson, 168

Rockefeller, David, 64, 70

Rumford, Byron, 53, 170n30

Rynerson, Jacqueline, 36–37

Saika, Peggy, 151, 152

San Francisco Bay: about, 1, 3, 4, 8–10; and bay development, 56–58, 62; and Berkeley Bay plan, 49–50; conclusion about, 162; conservation efforts for, 51; degradation of, 55; as a dumping ground, 19–20, 48, 69; and environmental legislation, 67, 69; garbage floating in, 48, 56; and Land Commission, 57; media attention for, 60–61; and multifamily dwellings, 111; private owners of, 55–56; report predicting condition of, 48; span of, 47–48; study group about, 59, 60

San Luis Obispo County, 75, 77

Santa Ana winds, 113, 133

Santa Monica Bay, 40, 131

Santa Monica Mountains: about, 2, 5, 8, 10; conclusion about, 161–62; and freeways, 121; homebuilding issues, 110; and large developments, 117–18; Los Angeles expansion into, 109; master plan for, 116, 117; and parks campaign, 119–21; and Philip Burton's bill, 127–28; and recreation area, 114, 123, 126, 128–31, 133, 161; turning point for, 125–26; women of, 123

Santa Monica Mountains Conservancy, 129, 130, 131

Santa Monica Mountains National Recreation Area (SMMNRA), 125, 126, 128, 130, 131, 132

Save Our Bay Action Committee (SOBAC), 67, 68

Save the Redwoods League, 20, 29, 53, 71

Save the San Francisco Bay Association, 1, 8, 50, 53–54, 59, 70

Save the Santa Monica Mountains, Parks, and Seashore, 2, 5, 124

The Saving of San Francisco Bay: A Report on Citizen Action and Regional Planning (Odell), 69

Saving the Bay: The Story of San Francisco Bay (series), 72

Scripps, Ellen Browning, 26

sequoias, 28, 29

Sessions, Kate, 25–26

sewage treatment, 41, 55, 146

Sherwood, Don, 61

Sierra Club: about, 2, 9, 166n18; "civil war" within, 91–94; and David Brower, 93–96; and Diablo Canyon, 92–97; Dunes campaign issue, 91–92; and environmental justice, 144, 156; first female president of, 12, 30; and Hetch Hetchy damming, 29; and Kathleen Goddard Jones, 81–82; oral history project, 100–101; sponsorship requirement by, 9; tax exemption issue, 65, 95; and Will Siri, 90

Silent Spring (Carson), 4, 21, 141, 159, 162

Silkwood, Karen, 140, 179n11

Siri, Will, 90
Slade, Afton, 15, 166n5
smog: about, 2, 160; alerts for, 14; campaign against, 14; impact of, 15; legislative session about, 15; and open spaces, 124; and women of SOS, 14–15
Sojourner Truth, 140, 180n12
solid waste management, 142, 180n18
Stamp Out Smog (SOS), 2, 13–15, 121, 138, 161
state parks, 29, 116, 119, 126, 128–29
Stringfellow Acid Pits, 9, 153, 154, 155
Sturgeon, Vern, 83, 90
suffrage movement, 6, 15, 28
Swift, Jill, 106, 122, 124–26, 129, 132, 161

tax revenue, 2, 76, 85, 91, 112, 114
toxic pollution, 146, 152, 153, 157
toxic sites, 145, 151
toxic waste, 9, 142, 147, 153

United Farm Workers, 8, 142
United Food and Commercial Workers union, 156–57
urban planning, 34, 51
U.S. Plywood Company, 114, 117, 118
utility companies, 3, 78

Varian Associates, 87
Vietnam War, 63, 64, 103, 137, 139
Voice of the River (Douglas), 22

waste facilities, 138, 141, 149, 180n18
waste products, 3, 19–20, 141, 153
water pollution, 35, 69, 159
Way, Howard, 68
Weaver, Harriett: about, 5–7; and brushfires prevention, 120; death of, 132; on first date, 107; handy skills of, 107–8; meeting with city council, 112–13; move to Los Angeles, 108–9; protests by, 113
Weaver, John, 6, 107
Westbay Community Associates, 64, 70
White, Laura Lyon, 3, 28
Wilson, Lee, 87–88, 90–91, 96, 99, 102–3
Women Strike for Peace (WSP), 22, 23
World War II: about, 2, 3, 4, 9; and Angel Island, 32–33; DDT use after, 21; postwar growth after, 108–10; suburban culture after, 17–18; women rejoining labor force after, 36

zero–population growth movement, 96, 143

Lightning Source UK Ltd.
Milton Keynes UK
UKHW010831290321
381018UK00014B/392